THE FORCE OF KNOWLEDGE

All measures of change which disregard the response of the human heart are either evil or naive.

Ivan Illich

THE FORCE OF KNOWLEDGE

THE SCIENTIFIC DIMENSION OF SOCIETY

JOHN ZIMAN, FRS

HENRY OVERTON WILLS PROFESSOR OF PHYSICS IN
THE UNIVERSITY OF BRISTOL

CAMBRIDGE UNIVERSITY PRESS

CAMBRIDGE

LONDON · NEW YORK · MELBOURNE

Published by the Syndics of the Cambridge University Press
The Pitt Building, Trumpington Street, Cambridge CB2 1RP
Bentley House, 200 Euston Road, London NW1 2DB
32 East 57th Street, New York, NY 10022, USA
296 Beaconsfield Parade, Middle Park, Melbourne 3206, Australia

First published 1976
Reprinted 1977

Printed in Great Britain
at the University Press, Cambridge

Library of Congress Cataloguing in Publication Data
Ziman, John M   1925–
The force of knowledge.
Includes index.
1. Science – Social aspects.   I. Title.
Q175.5.Z54   301.24′3   75–23529
ISBN 0 521 20649 9 hard covers
ISBN 0 521 09917 X paperback

CONTENTS

PREFACE

It was all very well for an enthusiastic bunch of leftists to conceive a Society for Social Responsibility in Science. After a few years, they converted this society to other ideological ends, and most of the reputable scientists resigned. But having confessed our manifold sins, and testified aloud for responsibility and relevance, we could hardly go back to our academic groves, to lecture, poker-faced, on the axiomatic foundations of quantum mechanics. Our own conscience, and our professional pride as university teachers, demanded that we introduce the theme of the social relations of science and technology to our students.

But how should this be done: by what method, in what form, in which department, by what persons, subject to what examinations? These questions are now being approached in a variety of ways by many different groups in many countries. It will take some years for a conventional curriculum to crystallize, and for this subject to become an ordinary academic discipline.

Meanwhile, and as a personal contribution to this development, I decided to have a go at it myself. This book is an enlarged version of a course of ten lectures given weekly for one term in the Faculty of Science at Bristol University from 1971 onwards. I thought that it would be more useful for one person to prepare and deliver these lectures, subject to all the follies of not being expert on any of the topics discussed, rather than calling in miscellaneous historians, economists, sociologists etc., each to say his specialized piece. You may describe it, if you like, as a teaching scientist's 'do-it-yourself' repair of his own defective education and wounded conscience, written down for the benefit of others facing the same problems.

Let me emphasize that none of the material presented here is very original: any good encyclopaedia or history of science will tell you far more on any particular topic. I had in mind the average second year Honours student of physics, chemistry or biology, who is very ignorant indeed of the barest facts or notions concerning science as a social activity. Such students, in my experience, are practical, realistic people, without deep interests in philosophy or sociology: they think better through concrete examples than through abstract theories. That is why I titivated the lectures with a large number of slides: at least it gave the audience something to look at whilst I was talking.

To make a success of a new 'subject', you have to examine in it. But I shuddered at the thought of setting 40-minute questions on this complex, confused, half-baked material. Do we really want them to *learn* who invented the zip-fastener, or the proportion of the GNP spent by the USA on R & D? To encourage attendance and interest in this subject, the Physics Department asked every student to write a substantial essay, during the Christmas or Easter vacation, on any topic in a very long list of appropriate titles. These essays were, in fact, very well done, and were carefully assessed for marks that count a small part towards the final degree result. I can only express a personal opinion that it would be a serious error to examine this new part of the science curriculum as if it were a cut-and-dried academic discipline with a well-defined list of facts and principles that every student should come to know – like the definition of an acid, say, or the experimental evidence for the wave nature of the electron. We teach and learn a few bits and pieces to get the feel of the subject, and to sensitize ourselves to other ways of thinking, not to acquire a machine for solving specific technical problems.

Apart from some helpful critical comments from Dr J. R. Ravetz, the text is wholly my own invention: but I am exceedingly grateful to Rosemary Fitzgerald who took on all the labour of finding the illustrations and acquiring glossy photographs and copyright permission. Not only does she know where to look, and whom to ask: she seemed to understand better than I did the sort of picture that was needed and to enjoy the sport of hunting for it in all sorts of out-of-the-way places. I am also most grateful to George Keene, who made all the slides for the lectures and took many of the photographs for the illustrations, and to Lilian Murphy, who interpreted my ugly hieroglyphics into an accurate typescript in her usual calm and careful manner. And I suppose I should be grateful for the British TV corporations, which kept my children fairly occupied, of an evening, whilst I tried to get on with this writing, and to my wife, who understood that this was a task that needed to be done.

JOHN ZIMAN

Bristol
July 1973

1 SCIENCE AS A SOCIAL INSTITUTION

Let us endeavour to see things as they are, and then enquire whether we ought to complain. Whether to see life as it is will give us much consolation, I know not; but the consolation which is drawn from truth, if any there be, is solid and durable; that which may be derived from error, must be, like its original, fallacious and fugitive.

Samuel Johnson

Natural science is transforming human society. In assessing the changes that science produces, we come to question the sources of its power. We have begun to doubt many opinions that were long taken for granted: that all science is good science; that scientific research is a way of personal purification; that support of science by the state is enlightened self-interest. Opponents of science, whose voices had long been muffled, have begun to say things that we thought would never be listened to again: that scientists are selfish, irresponsible and arrogant; that scientific knowledge is grossly misused; that mankind already knows too much for its own good.

This book is not a sermon on these moral issues. Each of us, as a responsible citizen of the world, must find his own answers to such controversial questions. But to think constructively about these matters, it is necessary to know a little bit about the nature of science as a human activity. It is not sufficient to understand the discoveries that scientists have made about the world; we must also learn to see scientific research as an integral part of the modern way of life. Rational debate on the political and moral issues concerning science and its place in society should be staged against some backcloth of facts and agreed principles. The aim of this book is to sketch out a background for such debates.

The trouble is (if I may say so) that the whole subject is much more complicated than many people realize. All too often, sweeping statements are made about the relationship of science to technology, or the proper machinery for the planning of research, or the wickedness of scientists when they do war research, without any regard to quite familiar facts that would contradict the whole argument. Various grand ideological schemes are put forward; a smoke screen of empty abstractions hides their failure to explain reality. Intellectuals of all parties ask for a 'deeper analysis' of the relationship between science and society without having explored and mapped out the surface of the subject. They tell us what should be done now, and what will happen next, without having looked at what really happened in the past, and where we now stand.

It is just not possible to gather together all the different current opinions on these topics, and to discuss their pros and cons. Nor can one set out a complete catalogue of the appropriate facts, from history, from philosophy, from politics, from economics, from sociology or from psychology; nobody is made wise by reading an encyclopaedia. I propose therefore to take up some of the significant themes that arise in these debates, and to illustrate them by reference to episodes chosen from history and from contemporary life. In each case I want to show the sort of evidence that might be used for, or against, a particular interpretation or general principle. In the spirit of natural science itself, I believe that one should try to acquire an overall impression of the relevant facts before trying to fit them into a theory.

The main effort has been to make every example as concrete as possible. Scientific knowledge, in its purest and most sublime form, is so much a product of the mind that we tend to ignore the body within which that mind must live. It is, of course, impossible for me to correct my own ideological prejudices in presenting this subject, but the deliberate bias has been sociological rather than philosophical. Scientific research should be observed as a daily task of particular people with a place in society; it should be seen as the organized labour of groups of people, banded together in social institutions such as universities and research laboratories, managing one another, paying one another salaries, and using expensive technical equipment. To emphasize the concreteness and reality of this way of life, the text is illustrated profusely with drawings, portraits, photographs, cartoons, graphs and tables of numbers – 'corroborative detail intended to lend an air of verisimilitude to an otherwise bald and unconvincing narrative'. As much can be learnt from these illustrations as from the text in which they are embedded.

The map of the intellectual world that is learnt by most science students looks rather like Fig. 1.1. The history of science, over many centuries, is represented as a continuous expansion at the expense of religion, philosophy and the humanities, which are left to scratch a meagre living

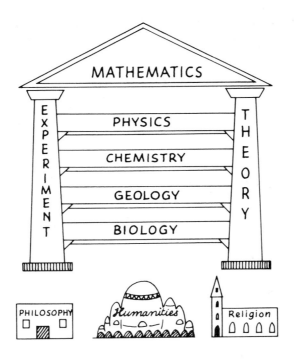

1.1

in a few barren corners. Some parts of these disciplines have begun to masquerade as 'social sciences', but they are only allowed this title when they speak in the language of formal theory and mathematical symbolism.

This naive and arrogant interpretation of the place of science in society will not survive a serious study of the facts. In Fig. 1.2 I have tried to indicate the complexity of the relationship between science and other human activities. In a general way *science* is taken to mean 'The Art of Knowing'. It is almost the same thing as *research*, which means the accumulation of knowledge by systematic *observation*, deliberate *experiment* and rational *theory*. But this activity is closely connected with the

3

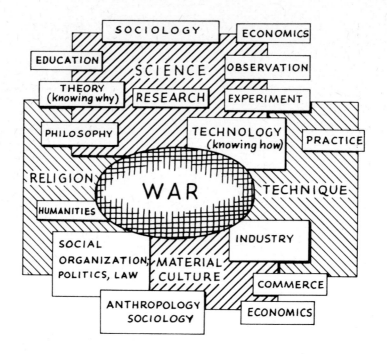

I.2

practical arts or *techniques* on the one hand, and with the spiritual sphere or *religion* on the other. These in turn link with one another in the *material culture* of Society, where they minister to individual human needs for food, health and psychological satisfaction.

But there are no sharp divisions between these different aspects of the human condition; each activity overlaps and merges with its neighbours. We all know the practical difficulty of drawing a line between science and *technology* – the 'Art of Knowing How' applied to an actual technique such as mechanical engineering or agriculture. How do we make a further distinction between the technique itself, and the technology that guides it? What is the subtle relationship between a technology such as medical science and the *practice* of a technique by experts such as physicians and surgeons? It is more correct to describe medicine as *both* a technique *and* a science – to say that it has theoretical, experimental, observational and practical aspects – than to force it into one or other of these boxes for the sake of mental tidiness. Rather than cluttering up the argument with pedantic definitions, let us allow the territories governed

4

by each of these terms to shade into one another and overlap without formal boundaries.

On the other side of the picture, science and religion are shown merging in an area that is also occupied by *philosophy* which draws especially heavily on *theory* – the special art of 'Knowing Why'. Here we make contact with the *spiritual needs* of men, such as are ministered to by *religion*, which is closely connected in its turn with the *social organization* that is required to provide for our *material needs*.

To lend a little more reality to our own map, it should be closed on itself, as if it were drawn on a cylinder or a globe. We need to bring *material culture* into contact with *science*, through *education, economics* and *sociology*. The latter belongs neither to technology nor to philosophy but bravely attempts, by observation and theory, to consolidate a body of knowledge concerning culture, religion and technique. Is sociology a genuine science? This is not merely a matter of definition or prejudice; claims to expertise, authority and power are at stake. Perhaps the ambiguity of the placing of these disciplines on our diagram is a fair indication of the uncertainty about where they really belong in the intellectual world.

Observe, at the very centre of our diagram, the most anti-human of human activities – *war*. Notice that it interacts with all other aspects of social life. I do not mean by this representation to give war the central place in modern society; but it has played, and continues to play, such an important part in the development of science and technique that it must not be banished to a distant corner of our minds, to be conveniently forgotten. Un-warlike science is almost as dated as un-scientific war, in our wicked world.

But this scheme is only an idle doodle, not meant to be taken literally. There are many other ways of analysing our subject. We could talk about the traditional hierarachy of the sciences, from the abstract mathematical properties of elementary particles, through atoms, molecules, cells and organisms, to the political behaviour of nations. We could make a political analysis, distinguishing carefully (and unfavourably) science under capitalism from science under socialism, and noting especially the injustices suffered by colonial nations under imperialist domination. It is sometimes convenient to see science as a point of balance along three dimensions of existence – the intellectual, the personal and the

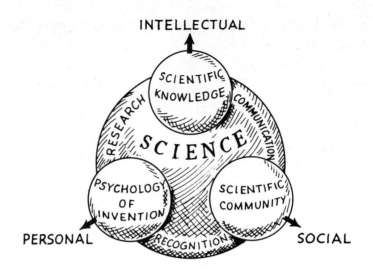

INTELLECTUAL

SCIENTIFIC KNOWLEDGE

SCIENCE

RESEARCH

COMMUNICATION

PSYCHOLOGY OF INVENTION

SCIENTIFIC COMMUNITY

RECOGNITION

PERSONAL

SOCIAL

1.3

social – all in tension with one another (Fig. 1.3). This diagram helps us to understand the complex relationship between the individual scientist and the scientific community. 'Scientific authority', for example, may be intellectual power acquired by research or it may be social standing as a recognized leader of the scientific community. The issues that may be depicted on this diagram are the special concern of the discipline now called the *sociology of science*, although obviously this is only a small portion of the whole subject.

In all human affairs, however, there is a single dominant variable – time. To make sense of the present state of science, we need to know how it got like that: we cannot avoid an *historical* account. In the language of physics, to extrapolate into the future we must look backwards a little into the past, so as to estimate the time derivatives of our functions. In the language of biology, there must be an *embryology* of science, explaining form through growth and growth through form.

But the detailed history of science is very subtle – and often very misleading. The deeper we pry, the less we see of pattern or principle. The further back we go, the more uncertain the facts, and the more ingenious their interpretation. It is a subject for the academic mind, giving more pleasure to the writer or lecturer than to the reader or student!

In the first half of the book, we look back over the past, often for many

6

centuries, choosing familiar and characteristic episodes to illustrate each theme. The main purpose is to demonstrate both continuity and change; in some of its features, such as the system of formal communication, science has scarcely changed since the seventeenth century; in other aspects, such as scale and internal organization, it has become entirely different within a lifetime. The proper contribution of history to sociology is to establish the time scales of change, and the variability of the circumstances in which social institutions may survive.

But these snippets of history are no substitute for a general knowledge of the actual development of the various sciences, such as may be obtained by any serious student, over a period of a few years, by systematic and miscellaneous reading. The examples are not important in themselves; they merely suggest what one could be looking for in such reading.

In the later chapters, we shall concentrate on the present century, on the past few decades, and on the present day, where (alas) it is our own fate to live. This material is not so readily available in academically digested gobbets, but one only has to look around inside laboratories, in the newspapers, and in a few specialized journals, to find plenty more evidence of the various phenomena to which reference will be made. To the sharp-minded student one may also recommend the practice of reading some of the more doctrinaire books on the theme of 'Science and Society' and seeking counter-examples to the generalizations that are confidently propounded by their authors.

Surprisingly, in a book with a sociological emphasis, much of the text deals with particular individuals. This is not because I believe that science is the activity of an élite, but because until very recently research was actually done by people working pretty much on their own, claiming *personal* rewards for their discoveries and exercising independent judgement on the problems to be tackled. It is precisely the transformation from this 'cottage industry/village market' system to the contemporary 'factory production/planned economy' style that is the main theme of this book.

2 WHICH CAME FIRST:
SCIENCE OR TECHNOLOGY?

The idea of the self-sufficient character of science ('science for science's sake') is naive; it confuses the *subjective passions* of the professional scientist, working in a system of profound division of labour, in conditions of a disjointed society, in which individual social functions are crystallised in a diversity of types, psychologies, passions (as Schiller says: 'Science is a goddess, not a milk cow'), with the objective *social role* of this kind of activity, as an activity of vast *practical* importance.   *N. I. Bukharin*

EARTH MEASURING

Every schoolboy learns the 'Theorem of Pythagoras', concerning the square of the hypotenuse of a right-angled triangle. As the name suggests it was supposed to have been discovered in about 500 BC by the famous Greek philosopher, or by some member of his School. The proof we learn is identical with that given by EUCLID of Alexandria, in his well-known compendium of geometrical theorems, written about 300 BC and transmitted to us through Muslim scholars of the Middle Ages (Fig. 2.1). But the same theorem proved in a slightly different manner (Fig. 2.2) was evidently known in ancient China (this figure shows an early example of the Chinese block printing, due to CHOU PEI, a contemporary of Pythagoras). Mathematics, surely, is the oldest and purest of sciences.

But the enormous pyramids and temples had been built, land had been surveyed and taxed, the heavens had been charted for the calendar, for several thousand years in Egypt and Mesopotamia. The practical use of the (3, 4, 5) triangle to construct a right angle must surely have been appreciated by the skilled stonemasons who used these tools (Fig. 2.3). The very word *geometry* means nothing more than 'earth measuring'. The intellectual feat of the Greek mathematicians in turning this practical art into a logical system must be admired; but the technique,

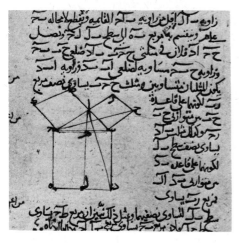

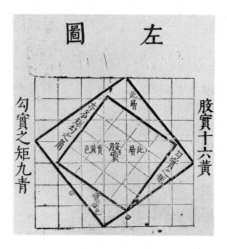

2.1 Theorem of Pythagoras from an Arabic text.

2.2 Chinese mathematical text. Early Chinese block printing. A vindication of the Pythagorean theorem, traditionally associated with the mathematician Chou Pei, probably a contemporary of Pythagoras.

with its rules discovered by experience, must have preceded the philosophical theory.

WEAPONS, TOOLS, TECHNIQUES

The extraordinary achievements of the Greeks in theoretical science and in the fine arts went along with an advanced technology. The obvious place to look for evidence of this is in military engineering.

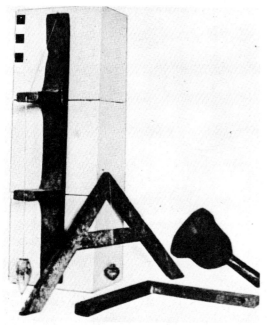

2.3 Egyptian stonemason's tools. Square level and plumb rule from the tomb of Sennedjem at Thebes, XX Dynasty: mallet from Saqqâra.

2.4   Reconstruction of a Greek siege weapon.

A siege weapon such as this (Fig. 2.4) was perhaps the largest and most carefully designed machine of its age. The most famous technical exploit of antiquity was the defence of Syracuse against the Romans in 215 BC. The great mathematician ARCHIMEDES (who despised such lowly arts, and published nothing on the subject) took charge of operations and used

2.5   Drawing of a fourth-century BC Greek galley.

his knowledge of levers and pulleys to design grapnels with which to hoist the enemy ships (Fig. 2.5) out of the water and destroy them.

But this was an isolated example of the foundation of a new technique upon scientific theory. The civilization of Rome persisted for centuries, and achieved marvels of technical skill, without any significant interest in theoretical science. Their capacity for fine and inventive craftsmanship in metal was unrivalled: a workman who could design and construct this lock and key (Fig. 2.6) could easily have made precision instruments

2.6   Roman lock. Model of one type of Roman lock and key, (above) shut, (below) open. The bolt slides in a pair of guides when drawn by the curiously cut key, whose projections fit slots cut through the bolt. To prevent the bolt from being slid without the key, it is held by 4 pins (*a*, *b*, *c*, *d*) pressed down by the spring into the slots, the pins being raised clear of the bolt when the proper key is inserted.

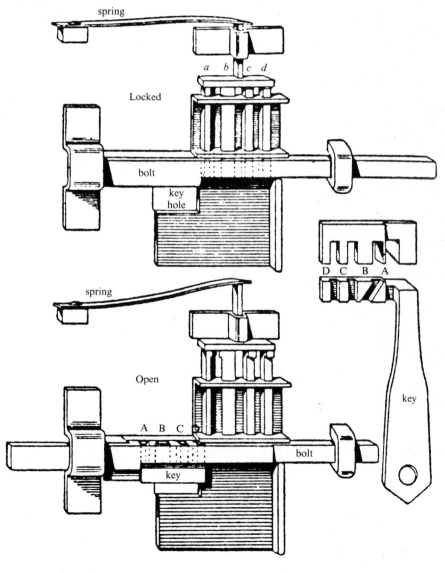

2.7 'Building the Tower of Babel' from the *Bedford Hours*, fifteenth century.

for use in navigation. But the Romans lacked some incentive to put astronomical theory to work for them. An advanced technique, learnt by apprenticeship and experience, is not necessarily consciously based on a body of abstract principles.

To appreciate the level of skill that may be achieved by the continuous development of a technique, we have only to look at the mediaeval cathedrals, whose design owed nothing to deliberate mathematical calculation of loads and stresses, and which were put together by illiterate craftsmen, under the direction of monkish architects, with crude wooden cranes and hand tools (Fig. 2.7).

Or look at this splendid machine for drawing heavy iron wire (Fig. 2.8). It was developed in the sixteenth century: yet see how efficiently and economically the man on the swing acts as the control unit

2.8   Mechanical wire-drawing. A water wheel drives a crank which on every half-turn enables the wire to be drawn through the draw plate. From Birinquccio, *De la Pirotechnica, 1540.*

connecting the power of the water wheel with the work to be done. As he swings forward he grasps the wire which is then drawn firmly through the die on the back stroke of the crank. A machine such as this is evidently the product of the inventive imagination of several generations of craftsmen, each improving a little on the technique he had inherited. Until the Renaissance, the existence and characteristics of such devices – wind and water mills, ships, mine gear, military engines etc. – lay outside the realm of interest of properly educated men, whose theoretical physics was dominated by subtle theological questions and garbled versions of Greek cosmology.

2.9 'An Alchemist at work.' Woodcut by H. Weiditz from Petrarch Trostspiegel, Augsburg, 1535.

ALCHEMY

But what did the alchemist, in this fifteenth-century woodcut (Fig. 2.9) think *he* was doing? Was he simply obsessed with a greedy dream of making gold – or did his researches have a higher philosophical purpose? Alchemy remains a mystery to us, for we cannot make up our minds whether it was sincere or fraudulent in intention. But notice the apparatus that he had made for his researches; the very prototypes of the modern chemist's flasks, beakers, retorts and stills. He had plenty of experimental and observational curiosity, and his research was guided by a systematic theory expressed in quasi-religious symbolism. Was it just an early case of a serious science that had got into a blind alley through faith in a confused and misleading theory, and could not rescue itself by open discussion and mutual criticism because it was a 'Secret Art'? Or was alchemy believed by its practitioners to be a practical technique, however wrapped up in mysterious mumbo-jumbo, that only failed to deliver the goods for accidental reasons? It has elements of true science and of genuine technique, and yet was neither completely scientific nor purely technical. There is a challenge here to any simple answers to the questions: 'How do we distinguish between science and technique: and which is primary?'

Those branches of science that used to be called 'natural history' rely heavily upon exploration of the earth and the observation of plants, animals, minerals and similar natural objects. This botanical collection (Fig. 2.10) carved on the walls of an Egyptian temple in −1450 was brought back from Syria by the Pharaoh THOTHMES III. Was curiosity the only motive for collecting specimens of unfamiliar plants? The

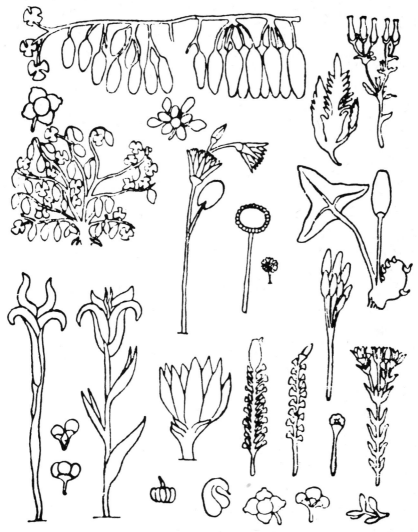

2.10 An early botanical collection. Strange plants and seeds brought back from Syria by Thothmes III, as they were carved on the walls of the temple at Karnak, Egypt, *c.* 1450 BC.

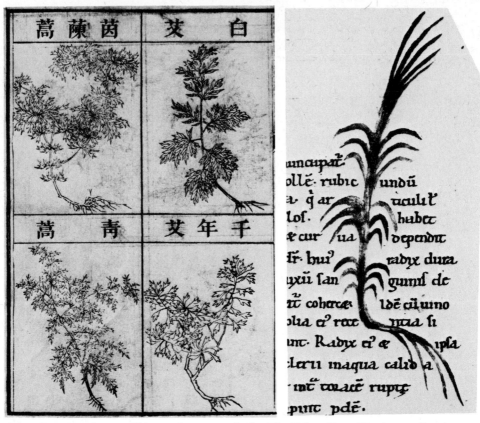

2.11   Drawings from a Chinese pharmacopeia, by Pên Tshao Kang Mu (AD 1596).

2.12   Herba Hipperis (*Equisetum* sp). The figure has probably been derived from a dead shoot of some Horsetail, perhaps *Equisetum arvense*. From *Herbal of Apuleius Barbarus*, twelfth-century manuscript.

drawings of plants from this Chinese Pharmacopoeia of about AD 1600 (Fig. 2.11) are perfectly accurate and recognizable. The top right-hand drawing is of *Artemisia alba*, which was recommended as a specific against intestinal worms. The efficacy of this drug for this purpose has since been confirmed by 'modern' medicine. Systematic botany thus begins, as in the mediaeval 'Herbal' (Fig. 2.12), from the highly practical technique of medical pharmacy; the physician needs to recognize accurately the plants from which he makes his remedies. The first botanic gardens (Fig. 2.13) were originally intended as basic material for the preparation of medicines, but soon proved their value as collections of plants that could be studied for their own sake. A scientific discipline arose out of a useful technology.

THE
HERBALL
OR GENERALL
Historie of
Plantes.

Gathered by John Gerarde
of London Master in
CHIRVRGERIE.

Imprinted at London by
Iohn Norton.
1597

2.13   Sixteenth-century herb garden. Frontispiece of *Gerard's Herbal*, London, John Norton 1597.

At this point, if I were attempting to write an outline history of science, I should begin to talk about the spirit of the Renaissance, the new humanism, the philosophy of Francis BACON (1561–1626), and the rise of the experimental method as an opponent of the 'book-learning' of mediaeval scholasticism. But this story is too well known, in various conflicting versions, to be repeated here. The question is: to what extent was the new natural philosophy of the seventeenth century connected with the techniques and technology of the day?

There is no doubt that the founders of the Royal Society wished to apply their new way of thinking to the improvement of the practical arts (Fig. 2.14). As shown by the influences on published research (Table

2.14 Page from Thomas Sprat,
History of The Royal Society, 1667.

ROYAL SOCIETY. 191

the obferving the apparent places of the Planets, with a *Telefcope* both by Sea and Land. This has been approv'd, and begun, feveral of the *Fellows* having their portions of the Heavens allotted to them.

5   They have recommended the advancing of the *Manufacture* of *Tapiftry* : the improving of *Silk making* : the propagating of *Saffron* : the melting of *Lead-Oar* with Pit-coal : the making Iron with Sea-coal : the ufing of the Duft of Black Lead inftead of
10 Oyl in Clocks : the making *Trials* on *Englifh* Earths, to fee if they will not yield fo fine a fubftance as *China*, for the perfecting of the Potters Art.

    They have *propounded*, and *undertaken* the comparing of feveral *Soyls*, and *Clays*, for the better making
15 of *Bricks*, and *Tiles* : the way of turning *Water* into *Earth* : the obferving of the growth of Pibbles in Waters: the making exact *Experiments* in the large *Florentine* Loadftone: the confideration of the *Bononian* Stone : the examining of the nature of *Petrifying Springs* : the ufing an *Umbrella Anchor* , to ftay
20 a Ship in a ftorm : the way of finding the *Longitude* of places by the *Moon* : the obfervation of the Tides about *Lundy*, the Southweft of *Ireland*, the *Bermoodas* , and divers parts of *Scotland* ; and in other Seas
25 and Rivers where the ebbing and flowing is found to be irregular.

    They have ftarted, and begun to practife the propagation of *Potatoes* ; the planting of *Verjuyce Grapes* in *England* ; the Chymical examination of *French*,
30 and *Englifh* Wines ; the gradual obfervation of the growth of *Plants*, from the firft fpot of life ; the increafing of *Timber*, and the planting of Fruit Trees ; which they have done by fpreading the Plants into many parts of the Nation , and by publifhing a large

Table 2.1. *Condensation of data on 'Approximate Degree of Social & Economic Influences upon the Selection of Scientific Problems by Members of the Royal Society of London'*[1]

Pure science	41%
Mining	21%
Marine transport	16%
Military technology	11%
Textile industry	3%
General technology and husbandry	8%

[1] From R. K. Merton, *Science, Technology & Society in Seventeenth Century England.* New York, Howard Fertig 1970.

2.1), the 'philosophers' and 'savants' of the seventeenth century concerned themselves very actively with all manner of projects of this kind, in military, naval, industrial and agricultural affairs. But they found the going much harder than they expected: except where sound theory already existed, as in astronomy and mechanics, little came of these efforts. As we shall see in later chapters, the task of putting a well-developed technique on to a 'scientific' basis is much more difficult than it appears at first sight. The most important and lasting achievements of the new philosophy were more fundamental, as in the Newtonian synthesis of mathematical physics or in the discovery of the microscopic world of biology.

THE TELESCOPE

Let us recall, however, one of the most famous experiments of 'pure' science of that era: here is NEWTON's record (Fig. 2.15), in his letters, of

2.15 Newton's prism experiment.

the discovery of the composite nature of white light; a sunbeam from a hole in the window blind is passed through a lens and a prism, which spreads it out into a rainbow on the wall. How is this 'pure' research connected with technology?

In the same year, 1671, NEWTON constructed his first telescope (Fig. 2.16), the ancestor of most of our modern astronomical telescopes. By

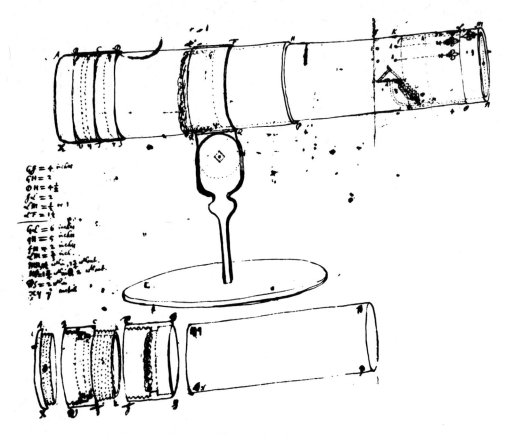

2.16   Newton's reflecting telescope – his own drawing.

that date the telescope was well-known as a scientific instrument. It is said to have been invented, more or less by accident, in 1608 by LIPPERSHAY, a Dutch spectacle maker. This 'curious toy' was immediately taken up by GALILEO (1564–1642) who greatly improved the design (Fig. 2.17), and used it to transform astronomy. But despite further improvements (based on the optical theories of KEPLER (1571–1630)) early seventeenth-century telescopes were not much use for

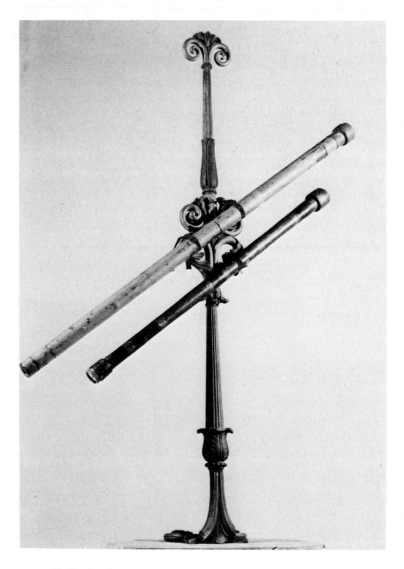

2.17　Galileo's telescope.

practical purposes. The trouble was that a simple system of lenses suffers
very severely from 'chromatic aberration' due to the differences of
refraction of light of different colours. The instrument had to be made
very long, with a narrow field of view. NEWTON's rainbow experiment
laid bare the source of the difficulty, which he was thus able to avoid by
using reflection from the curved mirrors instead of refraction through
lenses. Reflecting telescopes did not, in fact, come into common use; but
by the end of the century the trial and error methods of the commercial

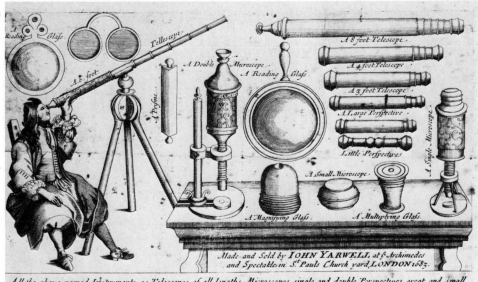

All the above named Instruments as Telescopes of all lengths, Microscopes single and double, Perspectives great and small Reding Glasses of all Sizes, Magnifying Glasses, Multiplying Glasses, Triangular Prisms, Speaking Trumpetts, Spectacles fitted to all ages, And all other sorts of Glasses, both Concave and Convex.

2.18 John Yarwell's trade card.

manufacturer (Fig. 2.18) and the application of the theoretical principles of optics produced instruments that could be used to some effect in war, on land and at sea. Within a few decades (1731) we find a compact telescope incorporated into HADLEY's octant (Fig. 2.19), which was a precision navigational instrument, accurate to within a fraction of a degree. International trade benefited enormously by a device that would allow one to find one's way about the oceans to within a mile or two, when previously one might make errors of landfall of hundreds of miles.

It had taken 120 years, from the invention of the telescope to the point

2.19 A Hadley octant in use.

where it had become of significant practical use. During this long gestation it had been through the hands of the pure scientists, and had come under the power of rational mathematical design. The story of other contributions to the technique of navigation – astronomical tables, logarithms, finding the longitude, the chronometer – consists of many similar sequences of scientific chickens and technological eggs.

STEAM POWER

The introduction of steam power, over a period of 200 years, is a familiar story (Fig. 2.20). Until the middle of the nineteenth century, this

2.20 The development of steam power.

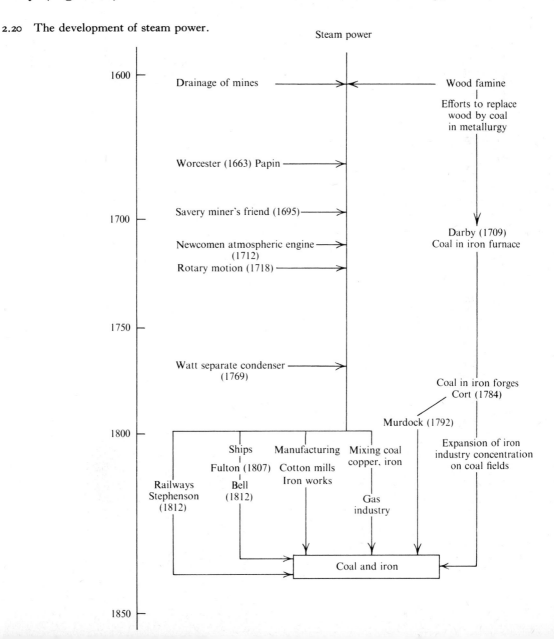

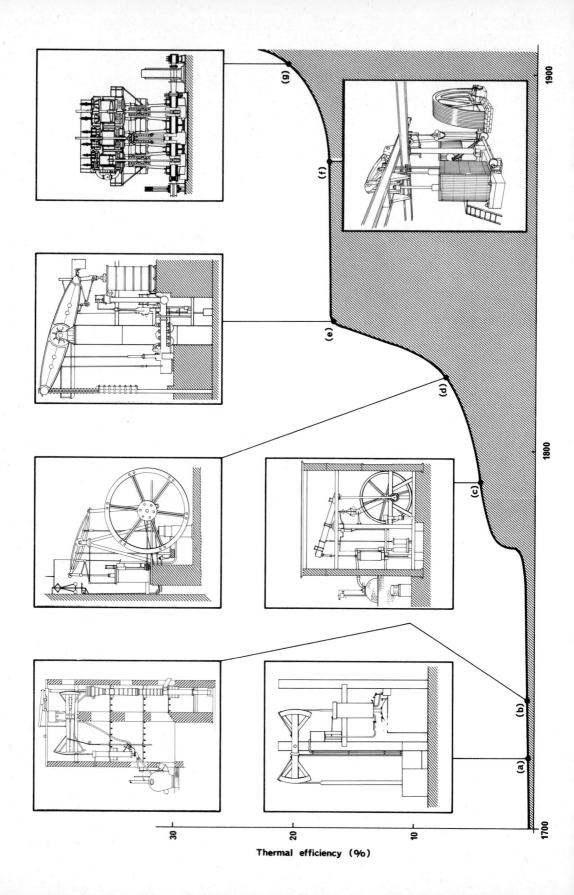

Thermal efficiency (%)

technical development – the most important in the whole history of modern industrial civilization – proceeded with scarcely any help from 'pure' science. The original incentive was strictly commercial and industrial – to solve the technical problem of pumping water from a mine. The only significant contribution from theory was the invention of the separate condenser in 1764 by James WATT (1736–1819) who had been an assistant to Joseph BLACK (1728–99), professor of natural philosophy at Glasgow University. BLACK's measurements on latent heat suggested to WATT the importance of avoiding the tremendous waste of heat in heating and cooling the cylinder of the Smeaton engine. Apart from this, the steam engine was invented and improved by a succession of practical inventors without any training in mathematics or physics. Indeed STEPHENSON (1781–1848) and TREVITHICK (1771–1833) were said to be 'scarcely literate'.

Nevertheless, quantitatively speaking, each step forward represented a substantial improvement in theoretical efficiency (Fig. 2.21). The engine manufacturers soon began to talk a language of simple economics, and by about 1800 would quote the 'duty' of each design – the number of millions of ft lb of water pumped up per bushel of coal burnt in the furnace. The most perfectly 'Marxian' quotation that I have ever seen in the history of science is the following, due to Thomas YOUNG (1807):

The daily work of a horse is equal to that of five or six men, the strength of a mule is equal to that of 3 or 4 men. The expense of keeping a horse is generally twice or thrice as great as the hire of a day labourer, so that the force of horses may be reckoned as half as expensive as that of men. On the authority of Mr. Boulton, a bushel (84 lbs) of coal is equivalent to the daily labour of 8⅓ men, or perhaps more; the value of this quantity of coal is seldom more than that of the work of a single labourer for a day, but the expense of machinery generally renders a steam-engine somewhat more than half as expensive as the number of horses for which it is substituted.

This is the sort of hard-faced calculation that led to the experiments of JOULE (1818–89) (Fig. 2.22) on the amount of heat that could be produced by a given amount of work, and hence, about 1850, to the

2.21 Improvement of thermal efficiency of steam engines. Data from Singer *et al. History of Technology*, Vol. IV, p. 164. (*a*) Newcomen (1718). (*b*) Smeaton (1734). Atmospheric beam pumping engine. (*c*) Watt (1792). (*d*) Woolf (1818). Compound engine. (*e*) Taylor's improved Cornish (1834). (*f*) Patent compound Corliss engine, by Messrs Douglas & Grant, Kirkcaldy (1877). (*g*) Triple expansion engine built for the Kiev power station of the Russische Elektrizitäts-Gesellschaft by Tosi of Legnano, Italy (1903).

2.22 Joule's apparatus for measuring the mechanical equivalent of heat.

formulation of the First Law of Thermodynamics – the principle of the conservation of energy. This law could easily have been derived as a simple consequence of NEWTON's laws of dynamics by the applied mathematicians and physicists, such as LAGRANGE (1736–1813) and LAPLACE (1749–1827), any time in the previous century: it took the cost accountancy of the engineers to bring it to light as a fundamental law of nature. Similarly, the Second Law of Thermodynamics arose historically from the cogitations of Sadi CARNOT (1796–1832) concerning the possible theoretical limits of efficiency of heat engines. After 1850 the general theory of thermodynamics could be used to design better steam engines – although we observe that the compound reciprocating engine of 1874 was not in fact more efficient than its parent of 1834. Until the middle of the nineteenth century, the steam engine did far more for pure science than science did for the technique of power engineering: it was not until near the end of that century that rational thermodynamic design could improve on experience, trial and error, and intuitive invention in this field of technology.

On the other hand, look at the history of electromagnetism. In 1820, in the course of a university lecture demonstration on the peculiar properties of current electricity, OERSTED (1777–1851) discovered the magnetic field produced by a steady current (Fig. 2.23). Within a few months a simple *galvanometer* for measuring current had been designed by SCHWEIGGER (1779–1857). This became the basis for many proposals

2.23 Lecturing before a group of select students, Oersted, in the spring of 1820, noticed that the needle of a nearby magnetic compass deviated when the circuit of a voltaic pile was completed. On 21 July the discovery of electromagnetism was announced.

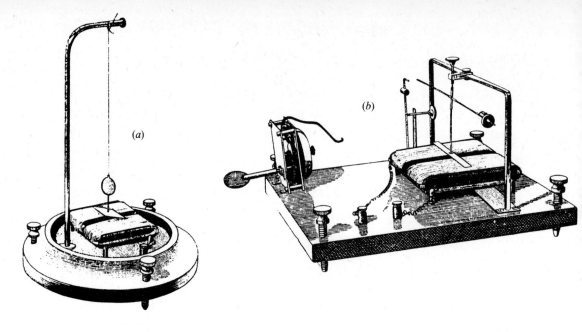

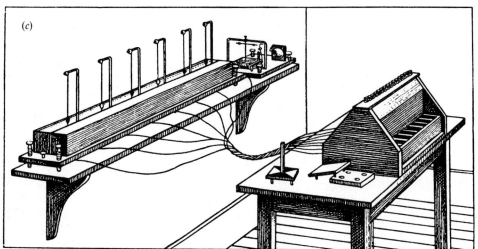

2.24 Schilling's telegraph *c.* 1825: (*a*) indicator, (*b*) alarm mechanism, (*c*) complete installation.

for a system of electric telegraphy: this design of 1825 (Fig. 2.24) by
SCHILLING (1786–1837) did not prove commercially viable, but by 1837
several inventors independently had developed practical systems. Here
is a clear case of an unforeseen discovery in the realm of 'natural
philosophy' being turned to social use. An interval of seventeen years
between the fundamental research discovery and its actual technical
application is not overlong in peacetime. A new technology of wires and

coils, batteries and switches had to be created, with new equipment to be designed and built, and new technicians to be trained. It takes time to change over from the laboratory scale and the experimental style to the industrial scale and the engineering style of thought.

In the case of electromagnetic induction, the delay between the scientific discovery and its full application was somewhat longer. FARADAY (1741–1867), the epitome of the pure scientist, was interested in the fundamental relationship between electricity and magnetism. In 1831 he discovered, during a series of deliberate researches with very simple apparatus (Fig. 2.25), that the motion of a magnet near a wire 'induces' an electric current. He immediately devised a 'magneto' – a primitive version of the electric generator which transforms mechanical power into

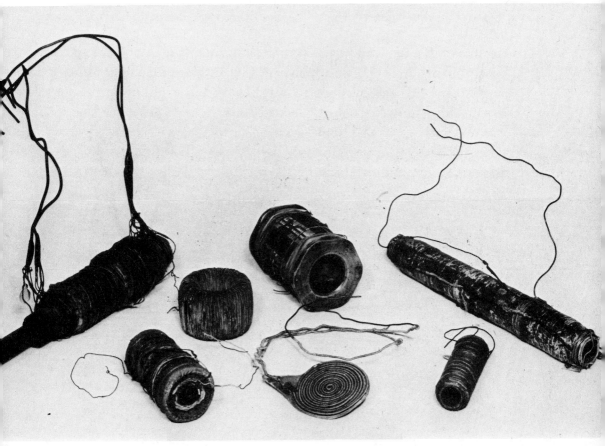

2.25 Faraday's electrical coils etc.

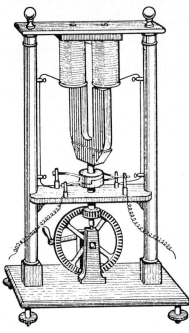

2.26 Hand-driven magneto electric generator of Hippolyte Pixii, *c.* 1833. The machine illustrated is fitted with a commutator of the type proposed by Ampère.

electricity in a modern power station. This one (Fig. 2.26) was actually made in Paris, in 1833, by Hippolyte PIXII. But these machines were used mainly for 'medical' purposes; it was believed that the shocks produced by the alternating current of 'electric fluid' were good for the health. It was not until about 1860 that big electric generators were constructed to power the electric arc lamps in lighthouses. The first public power stations were not built until the 1880s (Fig. 2.27) after the electric light

2.27 Nineteenth-century power station. Interior of Deptford power station under construction 1889.

bulb had been invented. We thus have a gap of fifty years between the basic scientific discovery and its technical exploitation on a large scale. In this interval the fundamental laws of electricity and magnetism were fully developed by academic physicists such as MAXWELL (1831–79), ready for the invention of radio at the end of the century. In the case of electromagnetism, therefore, theory was always far ahead of practice: the scientific stage came first, and owed almost nothing to social needs or practical techniques.

PLANT BREEDING AND GENETICS

As a final example, consider the technique of plant and animal breeding, an art which is as old as civilization itself. It is not easy to judge, from their descriptions, whether the varieties of melon listed in a modern catalogue are much superior to those advertised in 1873 (Fig. 2.28).

2.28   Seed catalogues: Suttons' 'Duke of Edinburgh' melon, 1873; catalogue of melons, 1973.

Sutton's Duke of Edinburgh. This is one of the largest and handsomest of the scarlet fleshed type, and can be especially recommended. It is the result of a cross between Orion and Scarlet Gem, possessing the strong robust constitution of the former with the excellent flavour of the latter. It is of oval shape, remarkably vigorous in growth, very free setter, grows to a large size (often 10 to 12 lbs. in weight), slightly ribbed, and beautifully netted. The flesh is a rich scarlet of extremely delicious flavour, and cannot fail to give complete satisfaction. A well known Melon grower informs us that it is the best of 25 sorts he grew last year.

Per packet, 3s. 6d.

Grow house varieties in a heated or cool greenhouse, sowing from February to May. Avoid shade and keep plants growing without check. Pollinate by picking mature male blooms, removing petals, and transferring pollen of male to female flower which has a tiny fruit at its base.

7111 SUTTONS EMERALD GEM (Green-fleshed).
Rich green colour flesh of unusual thickness. Superb flavour. *(See illustration)*   pkt. **17½p**

Some, indeed, of these varieties offered for sale in such glowing terms are merely modest improvements, under fancy names, of long-established stocks. In ordinary garden horticulture, the improvements of the past 150 years are scarcely more striking than those achieved in the previous era, from 1600 to 1800, during which commercial plant breeding became a recognized profession.

Breeding techniques, similarly, had not made great progress in that same period, until the last decade or so. By 1820 the basic methods of random, experimental hybridization were well established, with rule of thumb procedures to fix a new line after careful selection for desirable qualities. Detailed knowledge was being accumulated concerning, for

2.29   Darwin's pigeons. The English Pouter from Darwin: *Variation of Animals and Plants under domestication*, vol. 1, fig. 18, London, Murray 1868.

example, the pollination rules for apples or the types of characteristics inherited in the male and female line – the sort of lore that would be used even now, by the great-great-grandson of the Victorian founder of the firm, in his own laborious trial and error breeding programme to produce a new rose or cauliflower.

Remember, however, that the practical experience of animal and plant breeders in the early nineteenth century had its theoretical scientific consequences. An active period of observation and speculation on the mechanisms of inheritance culminated in the revolutions of thought that we associate with DARWIN (1809–82) and MENDEL (1822–84). DARWIN himself experimented with the breeding of pigeons (Fig. 2.29), realizing that the range of variability in natural populations was a key factor in the rate of evolution of new species. From MENDEL's numerical observations on the inheritance of specific characteristics in peas (Fig. 2.30) came the

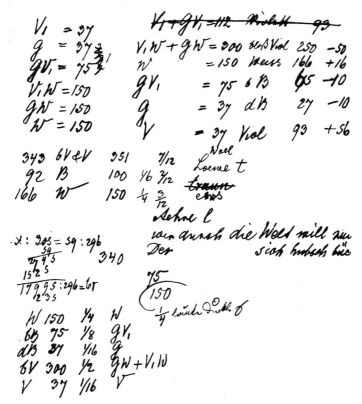

2.30 Mendel: the only remaining record of Mendel's experiments, known as 'Notizblatt'. According to R. A. Fisher it refers to the phenomenon of the recessive epistasis in *Phaseolus*.

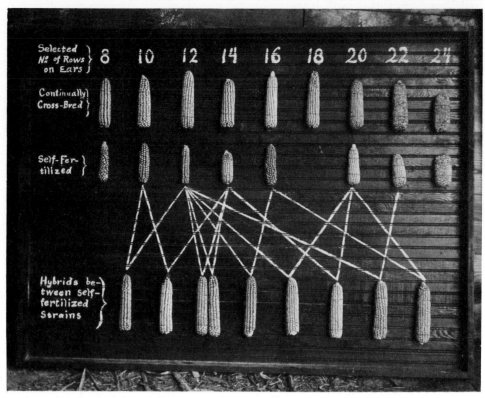

2.31 Dr G. H. Schull's first demonstration of heterosis in maize, 7 December 1909.

whole of our modern science of genetics, now almost as analytical and mathematically refined as theoretical physics. Since the 1950s this theory has been given its material underpinning by the discovery of the molecular structure of DNA, and of the genetic code. Here is pure science at its most subtle, refined and intellectually compelling.

But until quite recently this revolution of knowledge had little effect on the practical techniques. Here and there, in agricultural research laboratories and universities, there could be found plant breeders with sufficient grasp of Mendelian genetics and sufficient practical skill to marry the science with the art. This was an important factor in the long-drawn-out search for a new type of corn with hybrid vigour (Fig. 2.31), that produced such high yields when introduced into the US corn belt in the 1930s (see p. 263). Generally speaking, however, plant breeding made little direct use of genetic theory – the key fact, of course, in the rise of the charlatan T. D. LYSENKO to dictatorial power over practical and theoretical genetics in the Soviet Union during the Stalin era. It is only in the last few years that deliberate procedures, based upon detailed study of the genetic constitution of various varieties, have been used to 'engineer' valuable new combinations of qualities, such as 'dwarfness' or fungus resistance, into commercial crops. The 'Green Revolution' is a direct consequence of this technological revolution – and that may be only the beginning of a process of biological transformation of man and his environment. Over almost a century, from about 1850 onwards, we have seen the basic science of genetics separate itself from its parent techniques and spin around itself a cocoon of academic experimentation and abstract theory. Now, in our own time, this discipline has been reborn to the world of practice, and is generating a new technology of plant and animal breeding at a much higher level of predictability and power.

PRACTICE AND THEORY

There are many general theories concerning the historical relationships between science and technology. The examples I have given do not seem to conform to any simple theoretical model. Sometimes a technique precedes a science; at other times a new technology grows from a series of discoveries motivated by idle curiosity. Some techniques develop in close connection with parallel pure sciences; in other cases, practice and theory may separate for many, many years, and live almost independent lives, until they recombine fruitfully. According to Popper, scientific knowledge grows by the search for evidence to falsify imaginative hypotheses; in this chapter we have provided sufficient evidence to falsify almost all possible simple hypotheses of historical necessity. The lesson is, perhaps, that the history of science and technology is of sufficient diversity and richness that it cannot be summed up in an abstract formula. The examples that I have chosen are not more diverse in their historical and sociological structure than other cases that one can easily find for oneself by a little research.

3 WHO *WAS* A SCIENTIST?

A monarch may, by a single word, cause a palace to spring from the grass; but a society of men of letters is not the same thing as a gang of navvies.   *Diderot*

To understand the relationship between 'science' and 'society', we need to know something of the position of the *scientist* in the community at large. Who is he? What is he expected to do? To what social group, or rank, or class, or organization, or institution does he belong?

Here again, we need some feeling for history. The social 'role' of the scientist has evolved, over a period of many centuries, and some of the institutions of the scientific community are really very old. Many of the universities of Europe, for example, are as ancient as any of our political and religious organizations, and the scientific academies are much older than any of our industrial corporations.

But the task of describing the lives of scientists in detail would be quite beyond us. They were very *individual* people, who often left great masses of material about themselves and their discoveries. A modern scholar may make it his own life's work to mine the jewels to be found in the papers of a Newton or a Faraday. I have simply chosen three dates – 1670, 1770 and 1870 – and will try to look at some typical (if now very famous) scientists active at those times. A century, being somewhat longer than the life-span of an individual, is a good interval of time over which real social changes can be observed. In this chapter I shall not talk about their research work as such, but will try to describe them as they might have seemed to an ordinary contemporary. In the process we shall also see how they came to organize themselves into more self-conscious groups, as a coherent *scientific community* within the society of their day. This will be the background against which we shall study, in later chapters, the position of the scientist in the contemporary world.

36

3.1 Hon. Robert Boyle, FRS.
Portrait by J. Kerseboom.

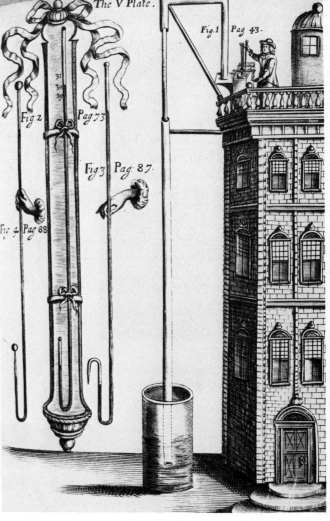

3.2 R. Boyle, *A Continuation of New Experiments Physico-Mechanical touching The Spring and Weight of the Air*, Oxford 1669.

SEVENTEENTH-CENTURY SCIENTISTS

Robert BOYLE (1627–91) (Fig. 3.1) is famous for his discoveries on the physics of gases, and for his treatise on chemistry, 'The Sceptical Chymist'. He was the fourteenth son of an Earl of Cork, and lived out his life in the aristocratic style of his day. You can see a corner of his great mansion in this diagram of his famous experiment (Fig. 3.2). He has been called 'The Father of Chemistry, and Uncle of the Earl of Cork!'. He was a child prodigy and a brilliant scholar, who devoted his last years to the intellectual defence of Christianity.

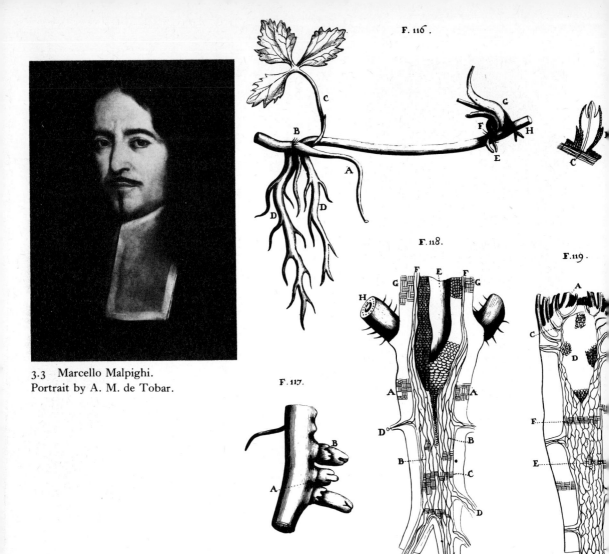

3.3 Marcello Malpighi.
Portrait by A. M. de Tobar.

3.4 Malpighi: microscopic studies of plants. From *Anatome Plantarum cui subjungitur Appendix heratas & auctas ejusdem Authoris de Ovo Incubato observationes continens*, tab. xxx in Appendix, London, John Martyn 1675.

Marcello MALPIGHI (1628–94) (Fig. 3.3) has been called 'the father of microscopy' for his numerous anatomical and botanical investigations with the newly invented microscope (Fig. 3.4). He was trained as a physician and spent much of his life as a professor of medicine at the University of Bologna, before becoming private physician to Pope Innocent XII in 1691. At that period, Italy was the leading scientific country of Europe, and the medical faculties of the Italian universities were already established centres of scholarly research (see chapter 6).

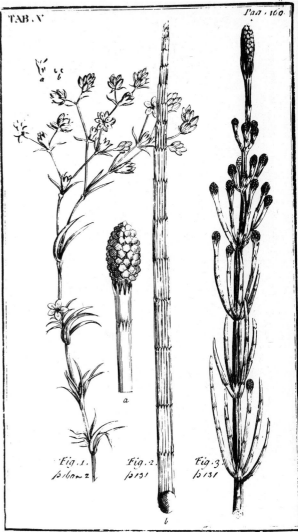

3.5 John Ray. Second frontispiece in *Stirpium Europaearum extra Brittannias nascentium Sylloge,* London, Sam Smith & Benj. Walford 1694.

3.6 Illustration of Horsetail from John Ray, *Synopsis methodica stirpium Britannicarum,* 3rd edn, London, Wm. & John Innys 1724.

John RAY (1628–1705) (Fig. 3.5) was the author of one of the earliest treatises in which plants are systematically classified (Fig. 3.6). He was the son of a blacksmith, but went to Cambridge and became a Fellow of Trinity College. In this post he had plenty of freedom to indulge his hobby of natural history – but in 1662 he was turned out of his fellowship because he refused to accept the religious tests of the Act of Uniformity. Fortunately, he was then supported by the patronage of a wealthy friend, and was able to continue his research.

3.7 Constantin Huygens and his family: Christiaan Huygens is the child above left.

3.8 Huygens' clock. Made by Salomon Coster, The Hague, 1657.

Christiaan HUYGENS (1629–95) lived in aristocratic comfort. He is shown here (Fig. 3.7) with his father, a distinguished diplomat whose name is better known in Holland as a poet than his son's as a scientist. But Huygens early made a reputation as a mathematician and astronomer – for example by building the first accurate pendulum clock (Fig. 3.8) – and was invited to Paris in 1666 to become a member of the Académie Royale. This was a paid post; but in 1681 he returned to Holland, perhaps because of the persecution of protestants by Louis XIV. He never married.

3.9 Medallion portrait of Leeuwenhoek in the frontispiece of A. van Leeuwenhoek, *Send-Brieven, Zoo ann de Hoog Edele Heeren van de Koninklyke Societieit to London*, Delft 1718.

3.10 Circulation of blood in the tail of an eel, demonstrated with one of Leeuwenhoek's microscopes.

Antonius LEEUWENHOEK (1632–1723) (Fig. 3.9) is one of the most attractive scientific figures of his time. He owned a smaller draper's shop and had little education, but his hobby was natural history. With a tiny globule of glass as a lens, he explored the microscopic world, and discovered unicellular organisms. To show their appreciation of his fame, the city of Delft made him honorary janitor of the town hall. Leeuwenhoek was thus enabled to indulge his hobby to the full and sent innumerable papers to the Royal Society of London reporting his observations. Here is the device he used to study the blood vessels in the tail of a living fish (Fig. 3.10).

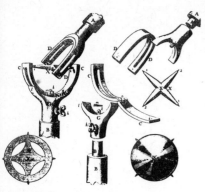

3.11 Robert Hooke's universal joint, c. 1676.

3.12 Hooke's weather scheme.

At one View repreſenting to the Eye the O[b]ſervations of the Weather for a Month.

Days of the Month and place of the Sun. Remarkable houſe.	Age and ſign of the Moon at Noon.	The Quarters of the Wind and its ſtrength.	The Degrees of Heat and Cold.	The Degrees of Dryneſs and Moyſture.	The Degrees of Preſſure.	The Faces or viſible appearances of the Sky.	The Notableſt Effects.	General [de]ductions [to] be made [af]ter the [...] is fitted w[ith] Obſervati[o]ns: As,
4 8 14 ♊ 4 12.46 12	27 12 ♉ 9. 45. Perigeū.	W. 2.9 3 12 3½ 16 10 W.SW.17	3 2 ½ 2 ½ ½ 2	5 29 8 9 29 29	¼ ⅛ ¾	Clear blew, but yellowiſh in the N.E. Clowded toward the S. Checker'd blew.	A great dew. Thunder, far to the South. A very great Tide.	From the quar: of the [N] o the chang[e] weather wa[s] [v]y temperat[e] cold for the ſon; the W[...] pretty con[...] between N. W.
15 ♊ 13. 40	28 8 ♉ 24. 51. 10	N.W. 3 9 4 N. 2 8 1 7	2 2 ½ 2	8½ 29 2 9 10 29	¼	A clear Sky all day, but a little checker'd at 4. P.M. at Sunſet red and hazy.	Not by much ſo big a Tide as yeſterday. Thunder in the North.	A little be[...] the laſt [...] Wind, and [...] the Wind ro[...] its higheſt [...]
16 ♊ 14. 37	10 N.Moon. S. at 7. 25 A.M. ♊ 10. 8.	1 10	1	10 28½		Overcaſt and very lowring.	No dew upon the ground, but very much upon Marble ſtones, &c.	Quickſilver [...] tinued deſce[n]ding till it c[...] [v]ery low; [...] which it be[...] o reaſcend[...]
	&c.	&c.	&c.	&c.	&c.	&c.	&c.	&c.

Robert HOOKE (1635–1703), on the other hand, was a man of many parts. He seems to have made his living chiefly as a surveyor and architect, being WREN's chief colleague in the rebuilding of London after the Great Fire. As research assistant to BOYLE, and then as full-time 'Curator of Experiments' to the Royal Society, he was enabled to prosecute a wide variety of researches in biology and physics. His microscopic discovery of the cells in plants is matched by such novel ideas as the design of a universal joint (Fig. 3.11), and a scheme for regular weather reports (Fig. 3.12). As a result of quarrels with NEWTON, his historical reputation is not very amiable, but in fact he was one of the cleverest and most sociable men of his day.

3.13 Isaac Newton: artist anon.

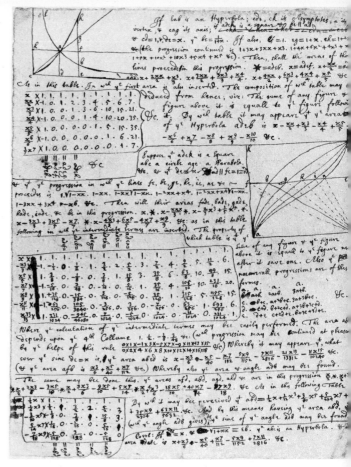

3.14 Newton's discovery of the binomial series by extrapolating a sequence in Wallis' *Arithmetica*. Winter 1664/5.

Isaac NEWTON (1642–1727) (Fig. 3.13) needs no introduction. His mathematical prowess early won him a Fellowship at Trinity College, Cambridge, and soon the Professorship of Geometry, for which the duties were negligible. What modern research professorship could offer to an absent-minded and solitary bachelor such a combination of domestic comfort and freedom to pursue his studies? (Fig. 3.14). He was very touchy and could not bear criticism, but after the publication of the *Principia* in 1687 he became exceedingly famous, and in 1695 was appointed Warden of the Mint where he did a thorough job of recoining the silver currency. For the last 25 years of his life, as Master of the Mint, President of the Royal Society, and with the title of Sir Isaac, he was regarded in England and in Europe as one of the greatest men of his time.

3.15 Gottfried Wilhelm Leibniz. English copy of an original German portrait.

3.16 Leibniz'
calculating machine.

Gottfried LEIBNIZ (1646–1716) (Fig. 3.15), NEWTON's rival as independent inventor of the Calculus, was his antithesis as a man of affairs. He was a child prodigy, a universal genius who spurned mere academic preferment in favour of diplomacy and high politics. As librarian to the court of Hanover, he served successive Dukes of Brunswick for 40 years, as much a senior civil servant as a famous scholar. His calculating machine (Fig. 3.16) greatly impressed the Royal Society when he visited London in 1673, but his contribution to philosophy surpasses even his mathematical work. Yet his death went almost unnoticed by his master, who had just become George I of England. He was, it seems, a *smooth* man – tolerant, urbane and perhaps a trifle crafty.

Quadrans Muralis Merid: 10 pedum Rad:

3.17   John Flamsteed. From an engraving by G. Vertue.

3.18   Flamsteed's mural arc at Greenwich.

John FLAMSTEED (1646–1719) (Fig. 3.17) was the first English Astronomer Royal. The post of court astronomer had been customary for centuries in other European countries (witness TYCHO de BRAHE (1546–1601) in Copenhagen and in Prague). CHARLES II got too good value for his money when he founded the Royal Observatory at Greenwich in 1676. FLAMSTEED was ill paid, but despite his poor health he laboured mightily on his own to build instruments such as the mural arc (Fig. 3.18), and to make an immense new catalogue of the stars. Like many other meticulous observers he did not get on well with the theorists; NEWTON treated him as a mere technician and wanted to use his data before he had reduced and checked them himself.

45

This is not meant to be an inclusive list. There were many more contributors to the scientific revolution in Europe alive and active in 1670. But it is still a small group, marginal to the main forces and institutions of society. A hundred or so names from Italy, France, Holland and Britain would cover most of the serious scientists of that era. To these we should add a number of dilettantes who liked to play at research – peering through telescopes or microscopes, collecting specimens of plants and shells, or experimenting vaguely in agriculture or medicine.

There were very few full-time posts for professional research: HOOKE and FLAMSTEED were ill paid, despite their great achievements. It was best to be born to wealth, like BOYLE and HUYGENS, or to find a patron, like RAY. It was almost incidental that professors such as NEWTON and MALPIGHI made such splendid scientific discoveries; in those days, active research was no more expected of a university teacher than it is now of a schoolmaster. Natural philosophy was essentially an obsessive hobby, in which a physician, a professor, a priest, a monk, an aristocrat or even a shopkeeper could indulge himself, just as nowadays he might take to rock-climbing or chess. In an age when there was genuine leisure for many members of the upper or middle class, research was almost entirely an amateur activity for a few well-educated or intellectually curious enthusiasts.

What was there about the social, economic, religious, political and/or philosophical climate of Europe in the seventeenth century that drew such people away from other private obsessions, such as cranky theology, mystical poetry, political intrigue or making money, into these new channels of thought and action? There is no satisfactory answer to this question. Some scholars have blamed capitalism; others discuss the effect of protestantism. Was it an essentially philosophical movement triggered off by the Renaissance and the Reformation, or was it geared to new techniques such as printing, mining and the ocean sailing ship? As a major problem in history or sociology, this question is of great academic interest, but any plausible answer would be almost irrelevant to the present state of science.

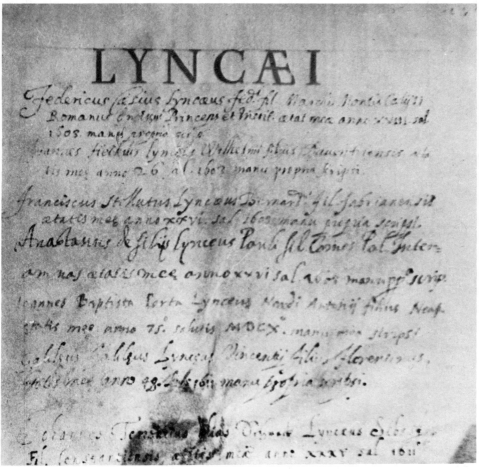

3.19 Act of Foundation of the Accademia dei Lincei, Rome, 1603.

THE SCIENTIFIC ACADEMIES

From this epoch, however, we inherit one of the most important institutions of the scientific community – the learned society or *scientific academy*. It was natural enough for the savants (as scientists were then called) to gather together in little clubs to discuss their research or to make experiments. The first properly constituted scientific society in Europe was probably the *Accademia dei Lincei*, founded in Italy in 1603 (Fig. 3.19), of which GALILEO was a member. This did not survive. But informal groups of scientists began to meet regularly, in Oxford, and in Paris, in the 1640s, and formed the nucleus of the *Royal Society of London* (1662) and of the *Académie des Sciences* (1666).

47

The frontispiece (Fig. 3.21) from Bishop SPRAT's *History of the Royal Society* puts it all in a nutshell. There is CHARLES II, the royal founder. It is not at all clear what his motives were in giving the Society a royal charter. Did his uncle, Prince RUPERT, a military technologist and considerable scholar, have any influence? It cost CHARLES almost nothing, anyway, and could have been thought to be useful on occasion. Lord BROUNCKER (1620–84), the first President, was a competent mathematician, and lent an aristocratic tone to the whole proceedings: the Royal Society has continued to belong to the Establishment, with adequate connections among the Top People. The most important figure represented is Francis BACON. Again, if we were to concern ourselves with the philosophical origins of European science, then Lord Chancellor BACON (who died in 1626) would be the hinge about which it all turned. His writings on experimental philosophy in the 1620s were the source of a new fashion in English thought, and the Royal Society was almost a realization of the 'House of Solomon' described in his *New Atlantis*. BACON argued, in the finest prose in English, that careful experiment and observation would infallibly lead to new scientific discoveries and to valuable technical advances. History has not entirely justified his optimism, for he underrated the role of formal theory, and did not appreciate the difficulty of improving existing practical methods by random experimentation. Nevertheless, the Royal Society was faithful to the programme implicit in Bacon's philosophy, and we do well to honour him as the prophet of scientific method.

The actual meetings of the Royal Society were devoted to the performance of experiments. The list given here by SPRAT (Fig. 3.20)

3.20 Experiments performed at meetings of the Royal Society.

Experiments of the Propagation of Sounds through common, rarify'd, and condens'd *Air*: of the congruity, or incongruity of *Air*, and its capacity to penetrate some Bodies, and not others: of generating *Air* by corrosive *Menstruums* out of fermenting Liquors, out of Water, and other Liquors, by heat, and by exhaustion: of the returning of such *Air* into the *Water* again: of the vanishing of *Air* into *Water* exhausted of *Air*: of the maintaining, and increasing a *Fire* by such *Airs*: of the fitness, and unfitness of such *Air* for respiration: of the use of *Air* in breathing.

Experiments of keeping Creatures many hours alive, by blowing into the *Lungs* with Bellows, after that all the *Thorax*, and *Abdomen* were open'd and cut away, and all the Intrails save *Heart*, and *Lungs* remov'd: of reviving *Chickens*, after they have been strangled, by blowing into their *Lungs*: to try how long a man can live, by expiring, and inspiring again the same *Air*: to try whether the *Air* so respired, might not by several means be purify'd, or renew'd: to prove that it is not the heat, nor the cold of this respired *Air*, that choaks.

3.21 Thomas Sprat, *History of The Royal Society*. Frontispiece to the 1667 edition.

(this is only one paragraph out of many pages) shows what a miscellany these were. To our eyes, they are rather absurd, but we must remember that the conventional wisdom of the day was a ragbag of old wives' tales, witchcraft, dogmas inherited from ARISTOTLE, and other mental bric-à-brac. Such research played an important part in clearing away superstition and establishing simple facts. The Society was also active in the collection of natural objects and books, organizing expeditions, manning official enquiries, and other useful business.

From our point of view, however, the Royal Society is really important as a meeting place for the scientific community. The savants were no longer isolated individuals; they belonged together in a recognized social group. As we shall see in chapter 5, the new academies immediately became centres for the *communication* of scientific knowledge. From this date, we may regard science as an *organized* social activity.

Despite its royal charter, the Royal Society was, from its foundation, a self-governing and self-perpetuating association without official power. Its funds were never lavish. Like any private club, it had to draw its income from the subscriptions of its Fellows, who carried no privilege beyond the right to the letters FRS after their names.

The Paris Academy, by contrast, had official status as an organ of the state. LOUIS XIV, in the manner of his government, made official visits (Fig. 3.22), paid pensions to its members, and provided funds for research. When it was reconstituted in 1699, the number of members was fixed, with three paid academicians in each of the fields of geometry, astronomy, mechanics, anatomy and chemistry. Each academician also had two associates and a pupil, in the true spirit of aristocracy and bureaucracy. The *Académie des Sciences* was thus much more like a government research institution than its English contemporary. The Berlin Academy, organized in 1700 by LEIBNIZ, who became its first president, followed a similar pattern. This difference between the English and Continental styles in the organization of science, reflecting the difference between parliamentary and autocratic government, still persists after 300 years. But the important principle that new academicians are elected by the existing members was established from the beginning.

3.22 Louis XIV and Colbert 'aux jardin du Roy' for a meeting of the Académie des Sciences. the background the Paris Observatory is under construction. Engraving after Sebastien Lecler

This list of scientists (Table 3.1) active in 1770, although very incomplete, is much too long to be studied in detail. The scientific community of Europe had grown steadily over the century, not merely in absolute numbers but by the spread of science to countries such as Switzerland, Sweden, Scotland and the United States. The internationalism of eighteenth-century science is apparent if we study the lives of men like LAGRANGE who served both the Berlin and Paris Academies, and EULER who was, in fact, the mainstay of the Imperial Academy of Sciences of St Petersburg, established by CATHERINE THE GREAT in 1725 on the French model. This institution, although of great distinction and a significant asset to Russian culture, was actually staffed almost entirely

Table 3.1. *List of 'natural philosophers'*

Bernouilli (D.)	1700–1782	Mathematics	Swiss	
Franklin	1706–1790	Electricity	American	Printer
Linnaeus	1707–1778	Natural history	Swedish	Professor
Euler	1707–1783	Mathematics	Swiss	Academician
Buffon	1707–1788	Natural history	French	Private means
Darwin (Erasmus)	1731–1802	Natural history	English	Physician
Haller	1708–1777	Physiology	Swiss	Physician
Hutton	1726–1797	Geology	Scottish	Manufacturer
Black	1728–1799	Chemistry	Scottish	Professor
Spallanzani	1729–1799	Physiology	Italian	Priest
Messier	1730–1817	Astronomy	French	
Lambert	1728–1777	Mathematics	German	Tutor
Cavendish	1731–1810	Physics	English	Nobleman
Priestley	1733–1804	Chemistry	English	Unitarian minister
Mesmer	1734–1815	Psychology	Austrian	Quack
Lagrange	1736–1813	Mathematics	French	Academician
Coulomb	1736–1806	Physics	French	Engineer
Galvani	1737–1798	Anatomy	Italian	Professor
Wolff	1733–1794	Embryology	German	Academician
Herschel	1738–1822	Astronomy	English	Music teacher
Scheele	1742–1786	Chemistry	Swedish	Apothecary
Banks	1743–1820	Botany	English	Private means
Haüy	1743–1822	Mineralogy	French	Priest
Lavoisier	1743–1794	Chemistry	French	Financier

by non-Russians, who were paid well to serve a foreign government. The analogy with the German refugees who made such a contribution to British and American science after 1930 is not quite correct; but the principle that scientific knowledge recognizes no frontiers is not new.

The continuity of scientific organizations through political revolution is remarkable. The Imperial Russian Academy was a body of professional research workers, more or less independent of the universities. The modern Soviet Academician is not only a distinguished scientist; he is also the paid director of a research laboratory under the control of the Academy. In the Soviet Union, and in other countries of Eastern Europe that have adopted the Soviet pattern, the universities are primarily teaching institutions, with very meagre facilities for research. This can be interpreted as a reflection of a more centralized and rationalized government system in communist countries – or is the difference from the practice of Western countries merely a product of the historical evolution of a traditional pattern in each major country?

By this time, in fact, the Royal Society (Fig. 3.23) had degenerated into a sort of fashionable club, open to distinguished people in general without any scientific claims. What, for example, could have been contributed to the learned proceedings by Lord BYRON? William SMITH

3.23 Men of science alive in 1807–8. Engraving by W. Walker & G. Zobel.

3.24 Portion of a geological map, by William Smith. From *A Delineation of the Strata of England and Wales...*, 1815. Detail of Plate XI.

(1769–1839), whose observations of geological strata and geological maps (Fig. 3.24) were the basis of all geological research, was only a canal surveyor; he was helped financially by the President of the Royal Society, Sir Joseph BANKS (1743–1820), but was not sufficiently genteel to be elected an FRS. BANKS had made his name as the botanist on James COOK's voyage to the South Seas in 1768 – an expedition organized by the Royal Society on behalf of the British government – but he was also an extremely wealthy man (see p. 135).

In addition to the academicians, the scientific community of 1770 also included a number of university professors. Some of these, like LINNAEUS, BLACK and GALVANI, actually taught the subjects on which they did research. But the curriculum in most European universities was still organized on the mediaeval pattern, emphasizing theology, classics and philosophy, and with professional schools in law and medicine. A chair of mathematics, natural history or anatomy provided suitable support for an occasional original scholar but there was very little encouragement of scientific research. To this day, most of the universities in Spain are similarly staffed by professors chosen for their skill in passing examinations, and for less relevant social aptitudes, without regard to ability in research; they are expected to devote themselves to lecturing at length on subjects whose permanence and continued relevance are taken for granted.

Science in the late eighteenth century was widely respected and officially encouraged. Nevertheless, most active scientists were still amateurs, with other means of support. It was an age of relative peace and prosperity where the leisurely life of the gentry, and of professional people like doctors and parsons allowed adequate time and facilities for private research. To play at chemical or electrical experiments, to have read one or two popular scientific works, to attend a conversazione at the Royal Society – these were fashionable hobbies, in keeping with the climate of opinion in the 'Age of Enlightenment'. It was also an age of great technical advance, in engineering and industry – but that was a different world, scarcely connected with the realm of natural philosophy.

SCIENCE IN THE NINETEENTH CENTURY

By 1870, the scientific community was quite transformed (Fig. 3.25). This is the century in which science became fully *academicized*.

3.25   Royal Society Conversazione in the library at Burlington House, 1908.

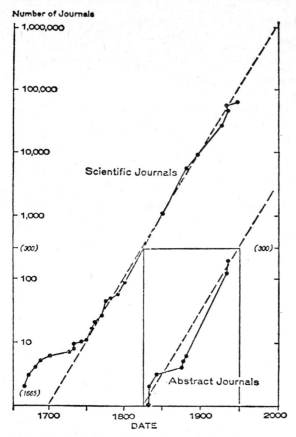

Number of Journals

1,000,000

100,000

10,000

Scientific Journals

1,000

(300)

100

10

(1665)

Abstract Journals

(300)

1700   1800   1900   2000
            DATE

3.26 Total number of scientific journals and abstract journals founded, as a function of date. Note that abstracts begin when the population of journals is *c.* 300. Numbers recorded here are for journals founded, rather than those surviving; for all periodicals containing any 'science' rather than for 'strictly scientific' journals. Tighter definitions reduce the absolute numbers by an order of magnitude, but the general trend remains constant for all definitions.

It had grown enormously. It is difficult to estimate the actual number of people contributing seriously to science in the latter half of the nineteenth century, but it must have been many thousands. The lists of famous scientists in the histories and bibliographies can no longer be expected to be complete. This is obvious from Derek de Solla Price's count of scientific journals (Fig. 3.26). If there were, indeed, something between 5000 and 10,000 learned journals in 1870, then there must surely have been at least the same number of readers and contributors.

This curve, showing the growth of scientific activity over the last three centuries, is one of the most important quantitative facts in this book. Plotted logarithmically, it fits very well to a simple law: *the 'size' of science*

56

has doubled steadily every 15 years. In a century this means a factor of 100. For every single scientific paper or for every single scientist in 1670, there were 100 in 1770, 10,000 in 1870 – and 1,000,000 in 1970. The extrapolation backwards is not quite accurate, and there are fluctuations about the main curve; but the general description is correct – with all its implications. Whether or not there will be 100,000,000 or so scientists in 2070, and 10,000,000,000 in 2170 remains to be seen; the rapid and uniform expansion from the past up to the present is the significant fact with which one must reckon.

It is impossible now to present a list of names, for individual discussion. We ought to have statistical analyses of the professional or social status of active scientists over the last century or so, but nobody has yet taken the trouble to make a head count.

ACADEMICIZATION

It is quite clear, however, that science had moved, lock, stock and barrel, into the universities. Most of the people who contributed to pure science were now engaged in academic work, as professors or would-be professors. A few wealthy amateurs, such as FIZEAU (1819–96) and DARWIN, were still indulging their hobby, and MENDEL, as everyone knows, was a monk. In medicine it was not unusual for a practising physician to experiment with new scientific techniques, and there were, of course, small numbers of government officials employed as experts in astronomy, geology, surveying etc. But to say that a man was a 'scientist' (the word was actually invented in 1840 by William WHEWELL) almost automatically implied that he had an academic post.

What were these professors of physics, chemistry, zoology etc. all doing? In England they were teaching students in a number of new universities that had started to offer science degrees. Until about 1850 it was not possible to receive a regular training in the experimental sciences. A rich young man might pick up a smattering by attending special lectures at the ancient universities. For a poor boy, such as Michael FARADAY (Fig. 3.27) the only entry to the scientific profession was by apprenticeship (1813) as a laboratory assistant. FARADAY was lucky: Humphry DAVY (1728–1829), the Director of the Royal Institution, recognized the brilliance of his assistant (see p. 29), and

57

3.27 Michael Faraday. Crayon drawing of Faraday in 1852, by George Richmond.

3.28 H. G. Wells as a student.

groomed him as his own successor. Towards the end of the nineteenth century, the degree of Bachelor of Science (or its equivalent, the 'Natural Sciences Tripos' at Cambridge) had become the regular qualification for research and for science teaching. Another poor boy, H. G. WELLS (Fig. 3.28) was enabled by a teacher's bursary to study biology at London University: but he failed his degree and instead of becoming a professor, he invented the Future. The natural sciences had now become 'subjects' at school and at university and gradually grew until they were being taught to an appreciable fraction of the youth of the nation (Fig. 3.29). In England, unfortunately, this took place by the creation of an optional 'science side', running in parallel with classical and 'modern' curricula in the higher forms of the Public and Grammar Schools. From this failure to bring science right into the older literary curriculum stems the deplorable present-day separation of arts and

58

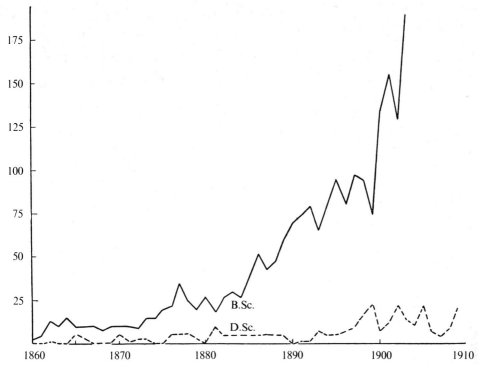

3.29 The number of successful candidates for science degrees in each year.

science into 'Two Cultures'. They order these matters better in most European countries, where scientific and humanistic subjects are combined, at school, up to the age of entry to the university.

GERMAN SCIENCE

But it was in nineteenth-century Germany that science became completely academicized. Many new universities were founded and began to compete for the ablest scholars, judged much more for their research output than for their capacity as teachers. For the first time, original scientific ability was considered the prime qualification for academic advancement.

Unfortunately, there were no paid posts below the rank of professor. A graduate trained in basic science and philosophy was expected to support himself as best he could whilst he prepared a dissertation for the senior degree of Doctor of Philosophy – so-called because that had been the

59

central discipline of the mediaeval university, rather like 'Arts' at Oxford and Cambridge. The aspiring scholar was then allowed to take pupils as a *Privat Dozent*, or 'Private Tutor'. But he also needed a private income, for he received no regular salary, even though he was expected to devote himself to research and teaching. The competition for preferment was intense, but if he was lucky he might eventually be called to a well-paid appointment as a professor. This plateau, achieved in his 30s or 40s, was comfortable enough and carried high social status, but there was still an incentive to continue research which might carry him into a more senior chair at a more famous university.

In many ways it was a harsh system, trading heavily on personal ambition and perseverance in the early years. But it transformed the style of scholarship in all branches of *Wissenschaft* – the German word that includes both natural science and the humanities. From the middle of the nineteenth century, German science leapt ahead, becoming much more rigorous, competitive and professional.

It also made research more cooperative. Each professor would acquire a group of assistants, constituting his *seminar* – what we should now call a 'research group'. These assistants were very dependent on his good will and patronage, for only the recommendation of your own professor could guarantee preferment to a chair elsewhere. Naturally enough the seminar became a 'school' of research, engaged in the solution of the problems proposed by the professor and devoted to his methods and scientific opinions. The university as a whole was little more than a federation of such groups, to which students would come for advanced lectures and research supervision. The connection between research and specialized teaching was thus firmly emphasized, at the expense of an integrated curriculum.

THE GRADUATE SCHOOL

This was, and has remained, the characteristic pattern of academic organization in Germany and other countries of Northern and Central Europe. In 1870 it was deliberately copied at Johns Hopkins University, and rapidly transformed the style of advanced scholarship at the major US universities. This is the origin of the modern *graduate school* where formal lecture courses and examinations are combined with a few years

3.30 The Inner court of the Cavendish Laboratory in 190

of quasi-original research, embodied in a written dissertation, for the Ph.D. degree. The German trend was resisted in England until the early years of this century, but has now become the standard form of advanced training in research throughout the English-speaking world. In France and Italy there were other traditions, but the actual succession of 'cycles' of degrees has very much the same content as in the American graduate school.

For the better part of a century, therefore, science has been closely connected with the universities. Nobody attempts to do research unless he has had at least an undergraduate education in his chosen discipline. Most research workers, whether or not they actually work in universities, have been through graduate school and learnt the technique of research on the way to the title of Ph.D. University teachers of science are expected to spend a great part of their time doing research, and the leaders of the scientific community, respected and honoured for their contributions to knowledge, are mainly employed as university professors. Even in an applied science such as medicine (see chapter 7), where professional practice plays such an important part, clinical research is largely concentrated in the teaching hospitals. It is hard for us now to imagine a world in which scientific research was not a major aspect of academic life, and the universities were not the main source of basic scientific knowledge.

As research became a more professional activity, it acquired its own special buildings. The Cavendish Laboratory at Cambridge (Fig. 3.30),

founded in 1871, by a private benefactor, was one of the first buildings dedicated specifically to research in pure science. The first Cavendish Professor, James CLERK MAXWELL, had no graduate students, but by the end of the century, under the leadership of J. J. THOMSON (1856–1940), this was one of the main centres of physics research in the world. Here, in 1895, came Ernest RUTHERFORD (1871–1937) from New Zealand, on a scholarship, to begin the researches on radioactivity that made him famous (see p. 215). In its new buildings, 'the Cavendish' is still one of the most important physics laboratories in the world (Fig. 3.31). The

3.31   Artist's impression of the new Cavendish Laboratory complex.

Institut Pasteur (Fig. 3.32) in Paris was established in 1888 around the great Louis PASTEUR (1822–95), who had discovered the role of bacteria in disease. The immediate purpose of this Institute was to deal with cases of hydrophobia, using Pasteur's technique, but medical research workers began soon to gather there until it became a world centre for the new sciences of bacteriology, microbiology and molecular biology.

THE SCIENTIFIC SOCIETIES

In our previous look at the Royal Society, at the end of the eighteenth century, we found it sinking into fashionable decadence: by 1870 it had completely reformed itself (Fig. 3.33). It was no longer open to mere

3.32 Popular image of Pasteur's laboratory at the Institut Pasteur in the *Journal Illustré*, 30 March 1884.

3.33 Burlington House, East Wing, built for the Royal Society in 1873.

wealth or high birth. To be elected an FRS was an honour signifying active research and positive contributions to science. The Fellows still included many amateurs, such as DARWIN and SORBY (see chapter 7), but the majority were professional academics. Victorian pundits like T. H. HUXLEY (1825–95) (see p. 137) used it as a base within the Establishment for organizing all the things that needed to be organized in education, in industry and in the government.

But there was (and remains to this day) a price to be paid for these high scientific standards and official influence. The business of the Royal Society was now cut off from contact with the lay public; it became rather esoteric and inward-looking, concerned mainly with the special affairs of the scientific and technical community. At the same time, through limitation in its numbers, it was too exclusive to act as an assembly of that community. Through the past century we may observe the failure of the Royal Society to expand at the same rate as the general scientific community, thus becoming more and more an aristocracy, or élite, rather than a body broadly representative of British science and technology. In many ways, the Royal Society of today is more self-regarding, and more concerned with maintaining its own status by careful selection of those whom it proposes to honour as Fellows, than it ever was in the past. The actual work that it does, by the administration of various funds and by advice on this and that, is rather insignificant by comparison with its inordinate pride in itself and the prestige of being a Fellow. It is much more like the House of Lords than the US Senate! Seeking influence rather than power, it is not corrupt but it is often ineffectual.

The failure of leadership in the nineteenth century is evidenced by the formation of new learned societies in the main branches of science (Table 3.2). Each of these began to do, in its own specialized field, all the things for which the older scientific academies had originally been founded. In the absence of adequate facilities for meetings and publications within the Royal Society, the geologists, the chemists, the zoologists etc. banded together and formed their own societies for the same purposes. To some extent, this was natural as a consequence of increasing specialization into separate 'disciplines' in the scientific and academic world. Yet it is noteworthy that this process did not occur, as it might have done, under the general umbrella of the Royal Society. The

Table 3.2. *Foundation dates of learned societies in London*

Royal College of Physicians	1518	Chemical Society	1841
Linnaean Society	1788	Institution of Mechanical Engineers	1847
Royal College of Surgeons	1800	Institution of Gas Engineers	1863
(successor to College of Barber-Surgeons)		Institution of Electrical Engineers	1871
Geological Society	1807	Physical Society	1874
Royal Astronomical Society	1820	Physiological Society	1876
Zoological Society	1826	Faraday Society	1903
Entomological Society	1833		

learned societies are extremely important in the modern scientific world as the means by which much of the practical work of publications, conferences, professional and educational consultation, etc. is carried out. An organization such as the American Chemical Society has an annual income of many millions of dollars, and publishes thousands of pages of scientific research. Neither in Britain nor the USA is there any general federation of such societies where the general business of the whole scientific community can be discussed and represented to the outside world. As for the French Academy of Sciences: its present state and future prospects are summed up in Fig. 3.34 which shows the evolution of the average age of election and age of death of Academicians, in the three centuries since its foundation!

3.34   Average age of entry (A) and exit (B) of members of the Académie des Sciences since its foundation.

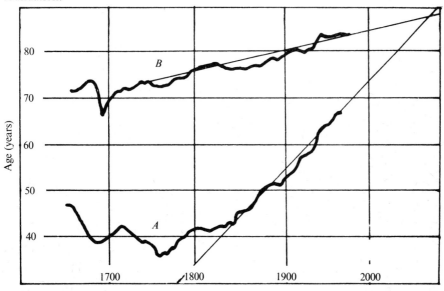

By the end of the nineteenth century, academic science was in full bloom; but industrial science was only pushing a few sprouts through the earth. Very little use was made of deliberate research in industry. Individual inventors such as Thomas EDISON (1847–1931) set up their own research workshops (Fig. 3.35) and applied scientific principles to the improvement of techniques in manufacture, agriculture, mining etc., but they seldom contributed to basic science. The modern idea of a research department in every factory had still to be born. Governments employed astronomers, geological surveyors, public health officials and other technical experts, but did not regard it as part of their duty to subsidize research into pure or applied science for general social

3.35 Edison's Machine Shop at Menlo Park, autumn 1879. The Edison dynamo (the 'long-waisted Mary Ann') in right foreground is attached by belts and pulleys to a dynamometer, also Edison's design, which measures the power produced by the steam engines in the room behind. Two field magnets lie beside the dynamo, and at the far left are several field electromagnet cores.

purposes. Dynamite was invented in 1866 by Alfred NOBEL (1833–96), who made a great fortune by manufacturing explosives; it was not the product of a research laboratory of any Ministry of Defence.

Nevertheless, the German dye and chemical industry prospered in the 1870s, and took the lead from Britain by the employment of professional chemists, not only for routine analytical testing but in the development of new processes and products. The universities produced many Ph.D.s with research experience: those who could not find places in the academic profession were welcome in industrial management and innovation. By the 1880s there were several firms in Germany and Switzerland employing a dozen or so scientists in their research laboratories, although these were still not on the scale of the university institutes presided over by such academic Grand Chams as von BAEYER (1835–1917) or HOFMANN (1818–92). Really large scale industrial research, about which we shall have much more to say in later chapters, belongs to our own time, from the First World War onwards.

FROM VOCATION TO PROFESSION

This chapter has had to be very sketchy. At every stage we could have digressed along many agreeable side alleys, noting the particular differences between one historical epoch and the next, and between one country and another. But the general picture is clear. Between 1670 and 1870 the scientist changed his position in the social framework. He begins as a very peculiar and isolated individual, a 'devotee' of science, noticed and honoured only if he makes some truly astounding discovery. By the end of the nineteenth century he has become a member of a recognized profession, ostensibly employed in academic teaching but encouraged and supported for his research. Through his learned societies and by his contributions to industrial 'progress', he has acquired high social status, and is regarded as an ornament and a valuable servant of his country. Yet he still works alone or as the pupil of an academic master, and largely chooses his own line of investigation. This is the historical basis of our present social position: these are the implicit assumptions concerning the relations between science and society by which many modern scientists still attempt to live. As we shall see, the present reality does not altogether conform to this genteel mythology.

4 STYLES OF RESEARCH

Everyone who is sure of his own mind, or proud of his office, or anxious about his duty, assumes a tragic mask. He deputes it to be himself and transfers to it almost all his vanity. While still alive and subject, like all existing things, to the undermining flux of his own substance, he has crystallized his soul into an idea, and more in pride than in sorrow he has offered up his life on the altar of the Muses.   *Santayana*

To be a scientist is to do research. This is not a self-evident activity, directed towards an obvious goal, like growing potatoes or flying an aeroplane. Research demands equipment and technique: it also demands an inner purpose. As we saw in the last chapter, the professionalism of modern science is of recent origin. Until our own generation, one did not take up research as if it were a job, like accountancy or architecture, from which a decent salary could be earned by honest toil. What were the motives that drove so many able men, over the past centuries, into this life of effort with so little material reward?

Explicit motives are seldom truly disclosed. Who knows why he does those things which he feels, most deeply, that he must certainly do? But we can look at the outside of a man's career, and ask what ends it seems to serve, and what style of life it represents. A few examples, from past and present, will quickly illustrate the richness and variety of the personal and intellectual modes within which first rate scientific work may be carried out. We need these examples, as a basis for discussion of the patterns of life now being imposed on the scientist, by society and by his own technical demands.

LEONARDO

The most famous scientific name of the Renaissance is that of LEONARDO DA VINCI (1452–1519). Not only was he one of the most celebrated artists of his day: he also devoted a considerable proportion of his very active life to research in many branches of science. In the numerous notebooks that have gradually come to light since his death are to be found innumerable original observations and theoretical ideas that were not to be rediscovered by others until years or centuries had elapsed.

This 'transparent torso' for example (Fig. 4.1), showing the internal organs realistically in their correct relationship, is only one of hundreds of anatomical drawings and sketches that LEONARDO made from his own dissections. By comparison all previous anatomical representations and descriptions are like the work of children. It was not until fifty years later that with the publication of VESALIUS' great treatise (see chapter 5) that anything approaching the same accuracy was achieved by others. Hidden in the same notes, we find the first observation of arteriosclerosis, the first representation of the cranal sinuses (the cavities in the skull), the idea of

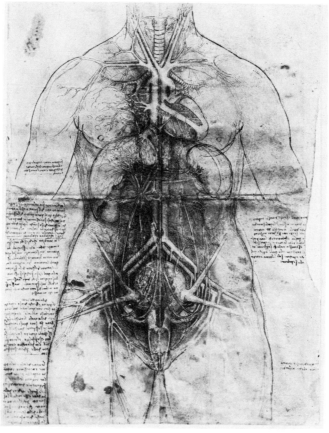

4.1   Leonardo da Vinci: 'transparent torso'.

studying the motion of the blood in a model of the heart, and so on. If he had been nothing but a crabbed student of human anatomy he would have been one of the greatest that ever lived.

LEONARDO's reputation as a mechanical engineer is familiar to everyone. Perhaps many of the ingenious machines that he sketched in his notebooks already existed in the mediaeval world – but his feel for the fundamental elements of a mechanism is obvious from these suggestions for various types of worm gear (Fig. 4.2). Some of the devices that he envisaged, such as conical roller bearings, were not reinvented until they were needed in our own century for delicate instruments such as the gyroscopic compass.

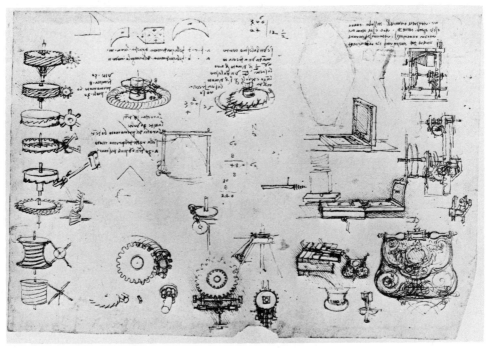

4.2   Leonardo da Vinci: some examples of gearing from his notebooks.

Again, in fluid mechanics, LEONARDO's drawings of vortex motion and waves (Fig. 4.3) are not only of great artistic beauty but are also informed with insight as to the forces at work. He saw that water waves transport motion but not matter, that vortices were formed at the widening of a canal, that the swiftest motion in a river is nearest the surface, and many other features of great theoretical importance.

4.3 Leonardo da Vinci: waterfall.

Throughout his work, indeed, he not only noted the facts but also asked fundamental questions which had never been asked before, and was concerned with the deeper meaning of everything that he studied.

Alas, as we all know, this marvellous intellectual effort went for nothing. His projected treatise on anatomy was never published, his inventions were never constructed, his scientific questions were not heard and could not be answered. From the point of view of the historian of science as a corporate enterprise of western man, LEONARDO might never have lived. Some small personal influences on other anatomists, engineers, physicists, botanists, or mathematicians is all that we can trace. As with his few great works of art, nothing was finished; all was in his own mind, his own sketches, his own private notebooks. The career of LEONARDO perfectly illustrates the fundamental role of *communication* in the scientific world (see chapter 5).

Yet it also illustrates research as a personal obsession. LEONARDO sought the impossible: he would encompass all knowledge of nature in its visual, realistic aspects. With a genius for observation and artistic representation he went deeper and deeper into every type of object that came his way – the human body, animals, plants, rocks, rivers, machines. His encyclopaedia could never be completed, for he was too restless, too undisciplined, too much of a perfectionist to stop at any point and set down for others a partial view. It is wrong to say that he was merely an artist of genius who happened to be interested in machinery and natural phenomena. LEONARDO's notebooks are records of the most intense research effort that has ever been made by one man alone, and are imbued throughout with the true spirit of *enquiry* that is the hallmark of modern science. In 1600 Francis BACON philosophized about science: a century earlier LEONARDO DA VINCI had already lived in the style that BACON advised.

CAVENDISH

Failure to publish important discoveries is a common characteristic amongst scientists. Henry CAVENDISH (1731–1810) was one of the greatest scientists of his time, and made many fundamental contributions to physics, chemistry and meteorology. He was the cousin of a Duke, and extremely wealthy, so was able to devote his whole life to scientific research. CAVENDISH was, indeed, an eccentric recluse, who never married and cut himself off from all female company. He seems to have been devoid of all sociability and all aesthetic sense, and lived a bleak narrow life with no distractions outside his research. This portrait of him (Fig. 4.4) was obtained by stealth, with the connivance of Sir Joseph BANKS, President of the Royal Society.

Within his own laboratory, however, CAVENDISH carried out some of the finest, most meticulous experimental researches on the fundamental problems of electricity and of chemistry. His most famous experiment was a direct measurement of the force of gravity acting between leaden balls hung on torsion wires (Fig. 4.5). More modern measurements of the same quantity have scarcely improved on the value he obtained nearly two centuries ago. But he would have earned his place in the history of science for his discovery that water could be decomposed

4.4 Henry Cavendish. Watercolour sketch by William Alexander.

4.5 Diagram of apparatus used by Cavendish to determine the density of the earth: *C*, outer casing; *PP′*, pulleys for rotating the beam *B*, from which 2 large balls (the weights), *WW′*, are suspended; *F* (broken line), inner casing protecting the torsion balance from draughts and temperature changes; *A*, thumb-screw for adjusting torsion balance; *l*, torsion wire; *r*, torsion rod, steadied by wires, *w*, supporting 2 balls *x, x′* at the ends; *LL′*, lamps; *TT′*, telescopes.

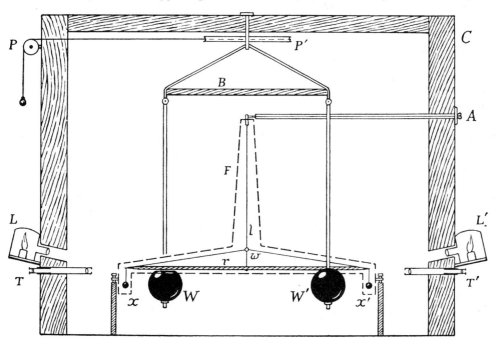

electrically into two gases, or for his elegant studies of the laws of electrostatics and of electrical conduction. CAVENDISH was not, I think, a conceptual innovator like FARADAY, nor a mathematical genius like NEWTON, but there is a precision and a finality about his research which puts him at the very top as a physicist. He saw science classically as the study of the *quantitative* aspects of nature and he dedicated his whole person to this demanding pursuit.

The fact that Cavendish only published a fraction of his research results does not mean, therefore, that he lacked application or self-discipline. In his notebooks he recorded and clearly stated such fundamental observations as the principle of chemical equivalence, the law of electrical resistance (i.e. 'Ohm's Law'), and the concepts of specific and latent heat, which were then unknown but which he did not write out for publication. It must have been, simply, that he cared very little about the rest of the world, and pursued scientific knowledge to please himself. Having observed, and elucidated a quantitative relation, he had satisfied his own curiosity and passed on to some other problem. Research was a private pastime, in which he engaged obsessively to satisfy his own peculiarly introverted personality.

RUMFORD

Cavendish's contemporary, Sir Benjamin THOMPSON, Count RUMFORD (1753–1814) was quite a different kettle of fish. He was no recluse, but he certainly cared little for his fellow men. A New England farm boy with no regular education, he raised himself by intrigue and treachery – marriage to wealthy widows, spying, political blackmail and other unscrupulous devices – to wealth, power and the title of a Count of the Holy Roman Empire. In youth he evidently had a winning way with his superiors: in later years he earned the hatred of his contemporaries by his overbearing and egotistical manner.

Yet this ambitious, unpleasant, self-centred man was one of the most 'socially responsible' scientists of his time. He is remembered now mainly for the experiment (Fig. 4.6) that demonstrated that heat cannot be a sort of fluid, but can be produced almost endlessly by frictional motion. This experiment grew out of RUMFORD's observation of the heat produced in the boring of a cannon, for he was then Minister of War to

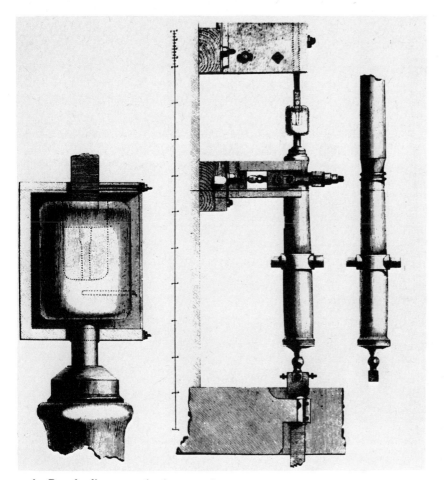

4.6 Rumford's cannon-boring experiment.

the Elector of Bavaria, and concerned himself particularly with equipping and training his army.

It was, indeed, precisely RUMFORD's cold calculating intellect, and love of efficiency and order that motivated much of his research. He wanted to clothe his men as cheaply as possible, so he began research on the conduction of heat through different types of cloth (Fig. 4.7). Puzzled by anomalies and failures of theory, he eventually discovered the phenomenon of heat transfer by convection and thus the now familiar explanation of the insulating power of materials such as fur where the air is trapped in innumerable small pockets. Similar motives determined his invention of the cooking stove, the steam radiator for central heating,

4.7 Rumford's thermometer for measuring heat conduction in liquids.

Scale of inches
0 1 2 3 4 5 6 7 8 9 10

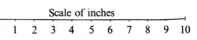

4.8　Count Rumford enjoying 'The comforts of a Rumford Stove', Gillray *c.* 1796.

greatly improved oil lamps, smokeless chimneys, etc. (Fig. 4.8). These simple inventions, which we now take almost for granted and which have certainly greatly benefited human existence, grew from his excellent grasp of the latest scientific principles, just as his more fundamental researches into the science of heat often had their origins in practical problems. It is difficult to discern a philosophical turn of mind in his research, but the simple ingenuity of his experiments and the real practical value of his inventions are evidence of an extraordinarily powerful intellect. He even invented the drip 'percolator' for making coffee (Fig. 4.9).

RUMFORD was, one might say, the forerunner of the beneficent *technocrat*, who enjoys putting mankind in order to satisfy his own ego. The 'eulogy' by Baron CUVIER (1769–1832) to the French Academy, on his death, puts this perfectly.

76

Nothing would have been wanting to his happiness had the amenity of his behavior equalled his ardor for public utility. But it must be acknowledged that he manifested in his conversation and in his whole conduct a feeling which must appear very extraordinary in a man so uniformly well-treated by others and who had himself done so much good. It was without loving or esteeming his fellow creatures that he had done them all these services. Apparently, the vile passions which he had observed in the wretches committed to his care or those other passions which his good fortune had excited among his rivals had soured him against human nature. Nor did he think that the care of their own welfare ought to be confided to men in common. That desire, which seems to them so natural, of examining how they are ruled, was in his eyes but a facetious product of false knowledge. He considered the Chinese government as the nearest to perfection, because in delivering up the people to the absolute power of men of knowledge alone, and in raising each of these in the hierarchy according to the degree of his knowledge, it made in some measure so many millions of hands the passive organs of the will of a few good heads. An empire such as he conceived would not have been more difficult for him to manage than his barracks and poorhouses. For this he trusted especially to the power of order. He called order the necessary auxiliary of genius, the only possible instrument of real good, and almost a subordinate divinity regulating this lower world. He himself in his person was in all imaginable points a model of order. His wants, his labors, and his pleasures were calculated like his experiments. In short, he permitted himself nothing superfluous, not even a step or a word, and it was in the strictest sense that he took the word superfluous. This was, no doubt, a sure means of devoting his whole strength to useful pursuits, but it could not make him an agreeable being in the society of his fellows. The world requires a little more freedom and is so constituted that a certain height of perfection often appears to it a defect, when the person does not take as much pains to conceal his knowledge as he has taken to acquire it.

4.9   Count Rumford's portable coffee maker.

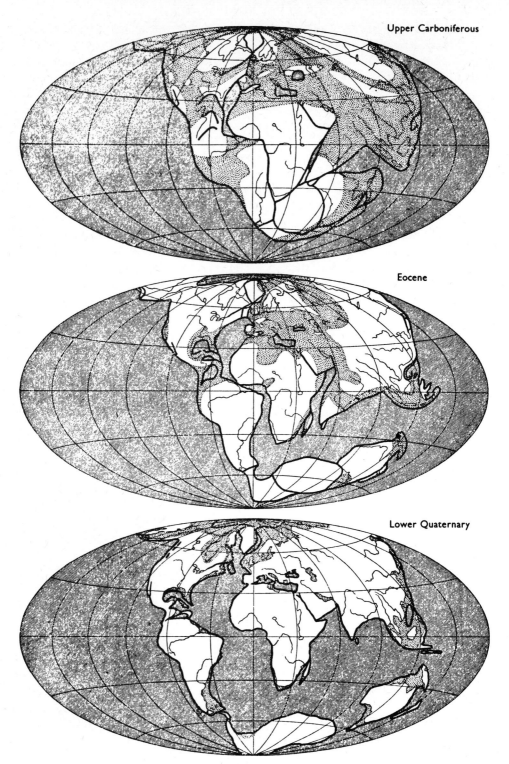

Upper Carboniferous

Eocene

Lower Quaternary

4.10 Reconstruction of the map of the world according to drift theory for three epochs. From Alfred Wegener, *Origin of Continents & Oceans.*

Alfred WEGENER (1880–1930) illustrates the crankiness of some research. In 1910 when he was a lecturer in meteorology and astronomy at the University of Marburg, in Germany, he was suddenly struck with a strange idea. For the rest of his life, until he was lost on an Arctic expedition in Greenland, he developed this idea in a succession of editions of a book entitled *The Origin of Continents and Oceans.*

The idea was simple enough – and not entirely original. The geographical fit between South America and Africa suggested that these great continents had once been one, and had split and drifted apart. The other continents could, more or less, be fitted together in the same sort of pattern (Fig. 4.10). Although not a geologist, WEGENER sought for evidence to support this hypothesis, from existing geological structures (Fig. 4.11), from palaeontological information about former climates,

4.11　Former relative position of South America and Africa, according to du Toit. From Wegener, *Origin of Continents & Oceans.*

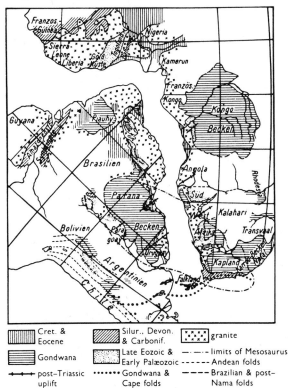

Cret. & Eocene	Silur., Devon. & Carbonif.	granite
Gondwana	Late Eozoic & Early Palæozoic	— ·—·— limits of Mesosaurus
		------- Andean folds
+--+--+ post–Triassic uplift	•••••• Gondwana & Cape folds	— — — Brazilian & post– Nama folds

from the distribution of species such as earthworms (Fig. 4.12), from earthquakes, and so on. He also gave geodetic evidence suggesting that Greenland had moved about one kilometre westwards in fifty years (Fig. 4.13), and proposed a mechanism for drift based upon forces produced by the rotation of the earth. His book, like his university lectures and indeed his whole personality, was direct, simple, open-minded and modest. His outward character certainly did not suggest a crank, arguing obsessively for a revolutionary theory in fields where he had no claim to expert knowledge.

Alas, although the theory of Continental Drift was widely debated in the 1920s, it was soon shown that the geodetic data were much too inaccurate and that the proposed dynamical forces were too small by a factor of a million or so. Other experts in palaeontology, tectonics, geophysics etc. found similar defects in WEGENER's arguments. As an opponent said of his research, 'it is not scientific but takes the familiar course of an initial idea, a selective search through the literature for corroborative evidence, ignoring most of the facts that are opposed to the idea, and ending in a state of auto-intoxication in which the subjective idea comes to be considered an objective fact'. After his death, WEGENER's heresy gradually dropped out of the teaching and textbooks

4.12   Present-day distribution of some earthworm genera of the family *Megascolecina*, superimposed on the pre-Jurassic reconstruction based on drift theory (according to Michaelson). From Wegener, *Origin of Continents & Oceans*.

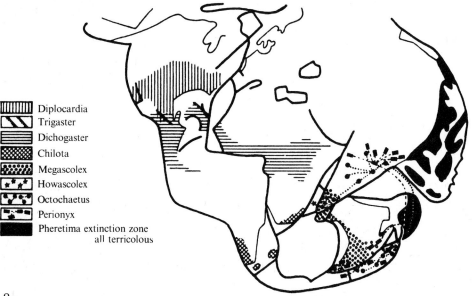

Diplocardia
Trigaster
Dichogaster
Chilota
Megascolex
Howascolex
Octochaetus
Perionyx
Pheretima extinction zone
all terricolous

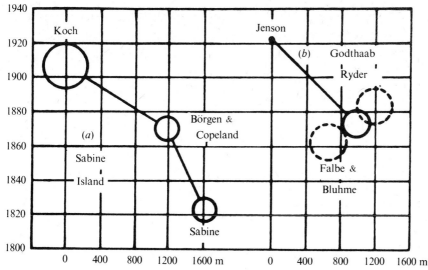

4.13 Displacement of Greenland according to earlier determinations of longitude. From Wegener, *Origin of Continents & Oceans*.

of geology. It is the duty of the scientific community to keep its minds, ears and books open to all sorts of speculative theories, however apparently far-fetched, and to give them a fair and serious hearing. It is also necessary to maintain the highest standards of intellectual criticism, and not to let mere conjectures, however superficially plausible in some respects, to be taken for truth. The danger of crankiness in research, as illustrated by the career of Alfred WEGENER, is quite as serious as the opposite folly of conservative opposition to new ideas.

Incidentally, the theory of Continental Drift is now considered to be perfectly correct.

RAMANUJAN

For perfect personal obsession, nothing can match pure mathematics. The brief life of Srinavasa RAMANUJAN (1887–1920) is a sweet sad fairy tale. His father was a petty clerk in a provincial town of Madras. He went through secondary school, but failed at the university and eventually got a minor clerical job. But when he was 15 he was lent a copy of a *Synopsis of Pure Mathematics* – not a textbook or treatise but an old-fashioned compendium of 6000 theorems with scarcely any proofs. RAMANUJAN devised his own proofs, and went on to state and prove numerous new theorems entirely independently of any other books or mathematical advice. These theorems were not merely schoolboy exercises. In a modest letter he submitted them to G. H. HARDY (1877–1947), Professor of Pure Mathematics at Cambridge, who was astonished at their subtlety. In ten years, working alone, RAMANUJAN had discovered many of the most difficult results obtained over the previous fifty years by the whole European mathematical community. In 1913 he was invited to Cambridge, became a Fellow of Trinity, and a Fellow of the Royal Society – but died of tuberculosis whilst still at the very height of his mathematical powers.

The moral of this is – presumably – that one cannot legislate for genius. RAMANUJAN's theorems are not, I guess, very useful at buttering parsnips, but they are very beautiful, and it is good that they should not be forgotten. He was, apparently, an agreeable mixture of the conventional Hindu clerk and the complete social outsider. Next time a slightly scruffy, rather plain little man shuffles apologetically round your door and says he has some interesting results concerning Goldbach's theorem, or on the transformation of space into time, or perhaps the quark model for elementary particles, it will be wise, for your own reputation, not to throw him out as a madman. He might conceivably have been Srinavasa RAMANUJAN, or Albert EINSTEIN.

WATSON

James WATSON (1928–   ), on the other hand, can only be himself. A clever young man with interests in ornithology and other aspects of biology, he found himself at the age of 22, with a Ph.D. of the University of Indiana, the first serious student of Salvador LURIA, an excellent geneticist. At that moment the search for a chemical mechanism for the inheritance of specific genes was nearing its climax and LURIA sent WATSON off to Copenhagen to learn biochemistry. With typical self-confidence, WATSON decided to move to the famous crystallography laboratory of Sir Lawrence BRAGG (1890–1971), at Cambridge, where he teamed up with an older and more experienced man, Francis CRICK (1916–   ). His book, *The Double Helix*, gives a racy, highly personal account of how Jim and Francis (Fig. 4.14) beat Maurice WILKINS (1916–   ) of King's College, London, and Linus PAULING (1901–   ) of

4.14 Watson and Crick at coffee.

Cal. Tech. to the solution of the structure of DNA, the vital molecule of heredity (Fig. 4.15).

CRICK and WATSON worked very hard on the problem; they got the personal credit for its solution and each won a Nobel Prize. But as with many other recipients of this award they had the advice and assistance of many other able scientific colleagues and contemporaries, this great achievement of the scientific community could only be recognized by a special award to the fortunate individuals who finally broke the last seals on the treasure, although a whole generation of geneticists, biochemists, microbiologists and crystallographers had contributed at various stages to the success of the enterprise.

The Double Helix is a fascinating book, not only for its journalistic exposé of the scientific life, but because it depicts an attitude towards research that is not uncommon in certain circles. The book reveals a young man preoccupied with his ambition to win a Nobel Prize and directing all his energies single-mindedly to that end. The DNA puzzle was, of course, difficult, and worth every bit of scientific effort to solve, but a reader is left with the impression that he chose this particular problem because of the fame it would bring him if he succeeded. The author seems preoccupied with the dreadful possibility that Linus PAULING, away there in California, one of the most brilliant chemists in

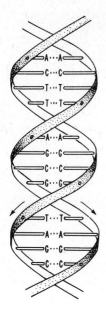

4.15 DNA molecule.

the world, might get there first. That would have been sad for Jim WATSON – but not in the least significant for the progress of knowledge.

FERMI

Enrico FERMI (1901–54) showed astonishing mathematical brilliance as a schoolboy, and quickly made his name as the outstanding student of physics at the University of Pisa. Within a few years, he had won an international reputation for his research in theoretical physics, and at the age of 26 became Professor of Physics at Rome. He owed a little to the patronage of an older man, Orso CORBINO (1876–1937), but this very early preferment was merely recognition of his complete mastery of all the physics of his day.

Enrico FERMI was the greatest all-rounder of his generation. He was a theoretician who could do beautiful experiments with his own hands. His personal intellectual power was matched with a modesty and simplicity that drew the best out of everyone who worked with him (Fig. 4.16). He was physically tireless, and could be gay company, yet preserved an inner gravity and reserve. He had an immensely quick and original mind, and knew all there was to know about physics, and yet could lecture and teach with utter clarity (Fig. 4.17). He was devoted to research, and cared nothing for fame or power, yet carried with full responsibility the

4.16 Enrico Fermi lecturing.

4.17 Enrico Fermi, *Notes on quantum mechanics.*

8-

hydrogen atom

(1) $\qquad U = -\dfrac{Ze^2}{r}$

(Neglect nuclear motion)
m will be reduced mass

Radial equation (7-7)

(2) $\quad v''(r) + \dfrac{2m}{\hbar^2}\left(E + \dfrac{Ze^2}{r} - \dfrac{\hbar^2}{2m}\dfrac{\ell(\ell+1)}{r^2}\right)v(r) = 0$

Put

(3) $\begin{cases} x = 2r/r_0 \qquad r_0 = \sqrt{\dfrac{\hbar^2}{2m|E|}} \\[2mm] A = \dfrac{Ze^2}{2r_0|E|} = \sqrt{\dfrac{mZ^2e^4}{2\hbar^2|E|}} \end{cases}$

(4) $\quad \dfrac{d^2v}{dx^2} + \left(\pm\dfrac{1}{4} + \dfrac{A}{x} - \dfrac{\ell(\ell+1)}{x^2}\right)v = 0$ $\quad \begin{cases} + \text{ for } E>0 \\ - \text{ for } E<0 \end{cases}$

Graphical discussion $g(x)$

$E < 0$

$v \to e^{-\frac{1}{2}x}$
and not $e^{+\frac{1}{2}x}$

therefore: adjustment required. Only discreet values of E allowable

$E > 0$

$v \to \dfrac{\sin}{\text{or} \cos}\left(\dfrac{x}{2}\right)$

no condition needed at $x \to \infty$
All $E>0$ allowable

Assume $E<0$ — Case of discreet e.values,

(5) $\quad \dfrac{d^2v}{dx^2} + \left(-\dfrac{1}{4} + \dfrac{A}{x} - \dfrac{\ell(\ell+1)}{x^2}\right)v = 0$

(6) $\quad v(x) = e^{-x/2}y(x)$

85

administrative burdens imposed upon him by the circumstances of his life.

Even in his political views, FERMI was the epitome of the modern scientist. In his younger days he was not interested in politics, and did not buck openly against the Fascism of Mussolini, who then ruled Italy and who showed him great favour. But the crimes of Hitler horrified him, and after emigrating to the United States in 1938 he enthusiastically played his part for his newly adopted country during the War: on 2 December 1942, FERMI's research team at the University of Chicago built the first nuclear reactor. From his experience in the government machine he became more concerned about the deeper political issues of freedom, and of war and peace; but he always remained mildly conservative, and relied more upon his own personal integrity and technical judgement than upon any ideological system. Perhaps, in the end, his central characteristic was that he would say nothing, whether about physics or about the world at large, that he had not thought out thoroughly and felt sure was right.

HALDANE

Yet one could not honestly say that Fermi was a more sincere, devoted and creative scientist than his contemporary, J. B. S. HALDANE (1892–1964), from whom he differed in almost every other respect. HALDANE, too, had a brilliant mathematical intellect and a protean memory, but his scientific contributions were in the fields of genetics and biochemistry. His father was a distinguished physiologist, who drew his clever son into his own research from childhood. From Eton and Oxford, HALDANE went into the First World War and fought with reckless, flamboyant, personal courage. Throughout his life he demonstrated his complete contempt for authority, and would pick on the most trivial issues to explode into frightful public rows with senior colleagues or administrators. This pugnacity was matched with a genius for public self-expression. He soon discovered that he had a gift for popular writing about science, and he became one of the best-known English scientists of his day, both for his vivid, simple, witty style and for the shocking frankness of the views he expressed.

In the Britain of the 1930s, there were many excellent scientists who

4.18   J. B. S. Haldane. Cartoon by David Low, *Evening Standard*, 2 December 1948.

took up with Marxism, in despair over the tragedies of unemployment and the crimes of fascism. But few were such active members of the Communist Party as HALDANE, who joined in 1938 and stood firmly by Stalinist policy throughout the War and into the Cold War era. For many years he was chairman of the editorial board of the party newspaper, the *Daily Worker*, where his popular scientific articles could be relied on to demonstrate the follies of capitalism and the superior virtues of a socialist viewpoint (Fig. 4.18). But he left the Communist Party around 1950, and in 1957 ostentatiously shook the dust of England off his heels to settle in India, thus dealing yet another blow at the aristocratic Establishment from whose roots he had sprung.

That was his public image, as seen, perhaps, through the distorting glass of the columns of the London *Times*. In private he could be kind and friendly, and was spoken of with love and devotion by many pupils and colleagues. His causes were often silly, his outrageous remarks often childish attempts to shock, and his quarrels with authority mere quixotic sallies without deep significance; but one could not help enjoying and admiring the wit, the learning, the hard sense and the perverse logic with which he would pulverize his opponents. Behind the *enfant terrible* lay a

87

shy and sensitive personality, for whom aggression was the only sure defence.

And he was a great scientist. He did not get a Nobel Prize (it is fascinating to speculate on his likely comment if he had!) but he was one of the first to show that Darwinian evolution and Mendelian genetics were mathematically consistent, and to calculate the rates of mutation necessary for the evolution of a new species. He applied these ideas to human genetics, and showed how a proportion of genetic defects associated with serious diseases such as haemophilia might arise. He made technical contributions to mathematical statistics, and was interested in cosmology, about which he often wrote for the general public. Our present views concerning the origins of life on earth were outlined by HALDANE in 1928, independently of the earlier work of OPARIN (1894–) which expressed similar ideas.

To get the flavour of HALDANE as a scientist, we should read about his physiological experiments, in which he was himself usually the experimental animal. Despite his public utterances in 1940 against the War (remember that Stalin was officially allied with Hitler at that tragic hour), HALDANE put himself and several of his closest friends into grave discomfort and very serious risk by a whole series of investigations on the effects of trying to live in confined spaces, under high pressures, with excess of carbon dioxide or low oxygen, not to mention cold and damp (Fig. 4.19). This research was the basis for much of the modern technique of underwater human activity, in submarines and scuba diving, and was of considerable importance in the War itself.

It is also interesting to study his reactions to the LYSENKO affair which was, of course, a direct assault upon the principles of his own genetic research, and which was, at the same time, supported as a dogma by STALIN and all the Stalinist parties. He expressed cautious support for LYSENKO (1898–) as a possible contributor to the scientific problem and was ready to propose that there might be a great deal in his theories and experiments. But as time went on, and as he saw himself being forced to set himself positively against 'Mendelism' he refused to compromise his scientific integrity, and quietly withdrew from official communist positions. It is easy to say that HALDANE should have been more forthright in defence, for example, of the personal position of VAVILOV (1887–1943), the great Russian geneticist, who was dismissed, imprisoned

4.19   J. B. S. Haldane.

and murdered by Stalin's police in their witch-hunt on behalf of
LYSENKO – but those were not perfectly easy times for a man who had
committed himself politically to that totally demanding cause. In his
enthusiasm for social responsibility as a scientist, he found himself, in
the end, with some pretty nasty bedfellows: it was not his form of
courage to jump out, denounce his former friends and say it was all a
mistake.

5 SCIENTIFIC COMMUNICATION

Nowadays, to be intelligible is to be found out   *Oscar Wilde*

Scientists form communities in other ways than through their learned societies or the universities in which they work. They are people linked together by their interest in particular scientific problems – the fracture of metals, pion-pion scattering, plate tectonics, photosynthesis etc. Those who belong together in such a group have been called an *Invisible College*, for theirs is a fellowship of the intellect rather than of material institutions or building. This term was revived by Derek de Solla Price; it was the name originally given to the scientific club that used to meet in Oxford in the 1640s, and which later became the Royal Society.

The connecting links between the members of such a group are not commands, or legal obligations, or financial transactions: they are bound together by the *communication* of information and knowledge. Science depends heavily upon the printed word for two reasons. It is essential to keep a permanent public record of results, observations, calculations, theories etc. for later reference by other scientists. It is also necessary to provide opportunities for criticism, refutation and further refinement of the supposed facts. In the elementary teaching of science we lay emphasis on making an accurate record of our experimental results and the conclusions to be drawn from them, and we learn to keep tidy laboratory notebooks. But the communication of the results of research to others is even more important. Science, by its very nature, is a body of *public* knowledge, to which each research worker makes his personal contribution, and which is corrected and clarified by mutual criticism. It is a corporate activity in which each of us builds upon the work of our predecessors, in competitive collaboration with our contemporaries.

The nature of the communication system is thus vital to science; it lies at the very heart of the 'scientific method'.

The actual forms of communication are varied, and have changed in emphasis over the centuries.

LETTERS

In the researches of the historian of science, the *private letter* remains one of the most important forms of evidence concerning discoveries and the diffusion of new ideas. This letter of 1671 (Fig. 5.1) from Isaac NEWTON

5.1 Letter of 16 March 1671 from Newton to Oldenburg.

to Henry OLDENBURG (1615–77), the Secretary of the Royal Society, tells of his recent progress with his new telescope (p. 20). It is typical of many such communications amongst colleagues, giving some notion of what one is doing, as between friends. It pretends to nothing more than a sharing of ideas, being written quite informally, without premeditation or redrafting.

The next example (Fig 5.2), however, again from NEWTON to OLDENBURG, is quite different. It was written in 1674; it is much longer; and it is in Latin. It represents, in fact, NEWTON's considered comments on a long letter by LEIBNIZ (see p. 44) concerning various algebraic series and is so carefully drafted that it could be published as it stood. Such a letter symbolizes the universality of scientific information. Latin was still a *lingua franca* for all educated men in Europe, and the symbolism of mathematics was also being conventionalized. Nowadays, a Chinese scientist writing to a Russian would use English (more correctly, perhaps, Broken English!) fortified with mathematical and chemical symbols, and biological nomenclature in Latin. The letter is carefully drafted because NEWTON and LEIBNIZ were more rivals than collaborators, and it would seem important to NEWTON to establish clearly his own results.

The continuing importance of 'private communications', as they are called in the scientific literature, is evident from this letter (Fig. 5.3) from James WATSON to Max DELBRÜCK (1906–) written on 12 May 1953. It explains the Crick and Watson model for the structure of DNA, which they had only just discovered (p. 83). It must be admitted, however, that there is a strong tendency nowadays to rush into public print with such communications. We even have special journals, with titles such as *Physical Review Letters*, where short 'letters' are published, reporting supposedly significant scientific discoveries, for the benefit of the whole scientific community. The creation of this type of journal is merely another stage in the process by which private letters – formerly the only means for the rapid transmission of scientific ideas – have been pushed into the public domain.

5.2   Letter of 24 October 1676 from Newton to Oldenburg. The 'Epistola Posterior'.

2.

variem, ut et area Hyperbolæ et caeterarum alternarum in hac serie $\overline{1+xx}\big|^{\frac{0}{2}}$. $\overline{1+xx}\big|^{\frac{1}{2}}$. $\overline{1+xx}\big|^{\frac{2}{2}}$. $\overline{1+xx}\big|^{\frac{3}{2}}$ &c. Et eadem est ratio intercalandi alias series idque per intervalla duorum pluriumve terminorum simul deficientium. Hic fuit primus meus ingressus in has meditationes: qui e memoria sane excidera nisi oculos in adversaria quædam ante paucas septimanas retulissem.

Ubi vero hoc didiceram mox considerabam terminos $\overline{1-xx}\big|^{\frac{0}{2}}$ $\overline{1-xx}\big|^{\frac{1}{2}}$. $\overline{1-xx}\big|^{\frac{2}{2}}$. $\overline{1-xx}\big|^{\frac{6}{2}}$ &c hoc est 1. $1-xx$. $1-2xx+x^4$. $1-3xx+3x^4-x^6$ &c eodem modo interpolari posse ac areas ab ipsis generatas: et ad hoc nihil aliud requiri quàm omissionem denominatorum 1, 3, 5, 7 &c in terminis exprimentibus areas; hoc est coefficientes terminorum quantitatis intercalandæ $\overline{1-xx}\big|^{\frac{1}{2}}$, vel $\overline{1-xx}\big|^{\frac{3}{2}}$, vel generaliter $\overline{1-xx}\big|^{m}$, prodire per continuam multiplicationem terminorum hujus seriei $m \times \frac{m-1}{2} \times \frac{m-2}{3} \times \frac{m-3}{4}$ &c. Ideo ut $\overline{1-xx}\big|^{\frac{1}{2}}$ valeret $1-\frac{1}{2}x^2-\frac{1}{8}x^4-\frac{1}{16}x^6$ &c Et $\overline{1-xx}\big|^{\frac{3}{2}}$ valeret $1-\frac{3}{2}xx+\frac{3}{8}x^4+\frac{1}{16}x^6$ &c Et $\overline{1-xx}\big|^{\frac{1}{3}}$ valeret $1-\frac{1}{3}xx-\frac{1}{9}x^4-\frac{5}{81}x^6$ &c. Sic itaque innotuit mihi generalis reductio radicalium in infinitas series per regulam illam quam posui initio epistolæ prioris antequam scirem extractionem radicum. Sed hac cognita non potuit altera me diu latere: nam ut probarem has operationes multiplicavi $1-\frac{1}{2}x^2-\frac{1}{8}x^4-\frac{1}{16}x^6$ &c in se, & factum est $1-xx$, terminis reliquis in infinitum evanescentibus per continuationem seriei. Atque ita $1-\frac{1}{3}xx-\frac{1}{9}x^4-\frac{5}{81}x^6$ &c bis in se ductum produxit etiam $1-xx$. Quod, & certa fuit harum conclusionum Demonstratio, sic me manu duxit ad tintandum e converso num hæ series quas sic constitit esse radices quantitatis $1-xx$ non possent inde extrahi more Arithmetico. Et res bene successit. Operationis forma in quadraticis radicibus hac erat.

$$1-xx\,\big(1-\tfrac{1}{2}xx-\tfrac{1}{8}x^4-\tfrac{1}{16}x^6 \text{ &c}$$
$$\underline{0-xx}$$
$$\underline{-xx+\tfrac{1}{4}x^4}$$
$$\underline{-\tfrac{1}{4}x^4}$$
$$-\tfrac{1}{4}x^4+\tfrac{1}{8}x^6+\tfrac{1}{64}x^8$$
$$\overline{0-\tfrac{1}{8}x^6-\tfrac{1}{64}x^8}$$

His perspectis neglexi penitus interpolationem serierum & has operationes tanquam fundamenta magis genuina solummodo adhibui. Nec latuit reductio per divisionem, res utique facilior. Sed et resolutionem affectarum aequationum mox aggressus sum easque obtinui. Unde simul ordinatim applicatæ, segmenta axium, aliasque quælibet rectæ ex arcubus datis vel innotuere. Nam regressio ad hoc nihil indigebat præter resolutionem aequationum quibus arcus ex datis rectis dabantur.

Eo tempore pestis ingruens coegit me hinc fugere et alia cogitans. Addidi tamen subinde condituram quandam Logarithmorum ex area Hyperbolæ, quam hic subjungo. Sit dFD Hyperbola cujus centrum C, vertex F, & quadratum inscriptæ $CAFE=1$. In CA cape AB, Ab, hinc inde $=\frac{1}{10}$ sive 0.1, & erectis perpendiculis BD, bd ad Hyperbolam terminati, erit semisumma spatiorum AD et $Ad =$ $0.1 + \frac{0.001}{3} + \frac{0.00001}{5} + \frac{0.0000001}{7}$ &c et semidifferentia $= \frac{0.01}{2} +$

UNIVERSITY OF CAMBRIDGE DEPARTMENT OF PHYSICS

TELEPHONE
CAMBRIDGE 55478

CAVENDISH LABORATORY
FREE SCHOOL LANE
CAMBRIDGE

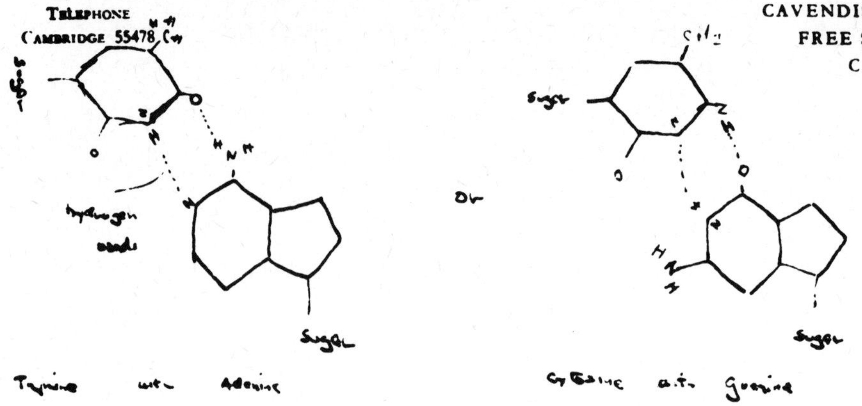

hydrogen bonds
Sugar
or
Sugar
Sugar
Thymine with Adenine
Cytosine with Guanine

While my diagram is crude, in fact these pairs form 2 very nice hydrogen bonds in which all of the angles are exactly right. This pairing is based on the effective existence of only one out of the two possible tautomeric forms – in all cases we prefer the keto form over the enol, and the amino over the imino. This is definitely an assumption but Jerry Donohue and Bill Cochran tell us that, for all organic molecules so far examined, the keto and amino forms are present in preference to the enol and imino possibilities.

The model has been derived almost entirely from stereochemical considerations with the only x-ray consideration being the spacing between the pair of bases 3.4 A which was originally found by Astbury. It tends to build itself with approximately 10 residues per turn in 34 A. The screw is right handed.

The x-ray pattern approximately agrees with the model, but since the photographs available to us are poor and meager (we have no photographs of our own and like Pauling must use Astbury's photographs) this agreement in no way constitutes a proof of our model. We are certainly a long way from proving its correctness. To do this we must obtain collaboration from the group at Kings College London who possess very excellent photographs of a crystalline phase in addition to rather good photographs of a paracrystalline phase. Our model has been made in reference to the paracrystalline form, and as yet we have no clear idea as to how these molecules can

5.3 Letter of 12 March 1953 from Watson to Delbrück.

Until the scientific revolution of the seventeenth century, the only way that new scientific ideas could be made public was through specially printed and published *books*. The great work of COPERNICUS (1473–1543), *De Revolutionibus Orbium Coelestium* (*The revolution of the heavenly spheres*) (Fig. 5.4), appeared in the week of his death in 1543, and sold very few copies. But his ideas were spread around amongst the few professional astronomers, even if they scarcely reached the general

NICOLAI CO
PERNICI TORINENSIS
DE REVOLVTIONIBVS ORBI-
um cœleſtium, Libri VI.

Habes in hoc opere iam recens nato,& ædito, ſtudioſe lector, Motus ſtellarum , tam fixarum, quàm erraticarum, cum ex ueteribus, tum etiam ex recentibus obſeruationibus reſtitutos:& no- uis inſuper ac admirabilibus hypotheſibus or- natos. Habes etiam Tabulas expeditiſsimas , ex quibus eoſdem ad quoduis tempus quàm facilli me calculare poteris.Igitur eme,lege,fruere.

Ἀγεωμέτρητος οὐδεὶς εἰσίτω.

Norimbergæ apud Ioh. Petreium,
Anno   M.   D.   XLIII.

5.4   Title page of *De Revolutionibus Orbium Coelestium*, Nuremburg 1543.

5.5　Frontispiece of Vesalius' *Fabrica*, Basel 1543. (Attributed to Jan Stevenszoon van Calcar.)

public. The importance of this book was not realized until well into the seventeenth century, when Kepler brought it to the attention of all serious astronomers and GALILEO made it the focal point of his struggle for a new philosophy of nature. On the other hand, *De Corporis Humani Fabrica* (*On the structure of the human body*) (Fig. 5.5) by Andreas VESALIUS (1514–64), which was published in the same year, caused an immediate impact, and was widely published. For centuries it retained its appeal and function in the teaching of medicine – not only as the very *first* accurate book on human anatomy with illustrations, but as the finest, most beautifully drawn collection of such illustrations that was ever published. Remember, however, that LEONARDO (p. 68) had made many such drawings fifty years before – but had lacked the systematic patience and technical medical training to make a complete work. In the book of anatomy, as in the mediaeval or renaissance *herbal* (Fig. 5.6), we see the printed work as something more than an announcement of a new scientific discovery: it is an indispensable tool of practice and research, comparable with the modern computer or microscope. The coincidence of the rise of science with the discovery of printing is not at all an accident.

5.6 A mediaeval herbalist in his study, by Adrian van Ostade.

ON

THE ORIGIN OF SPECIES

BY MEANS OF NATURAL SELECTION,

OR THE

PRESERVATION OF FAVOURED RACES IN THE STRUGGLE
FOR LIFE.

By CHARLES DARWIN, M.A.,

FELLOW OF THE ROYAL, GEOLOGICAL, LINNÆAN, ETC., SOCIETIES;
AUTHOR OF 'JOURNAL OF RESEARCHES DURING H. M. S. BEAGLE'S VOYAGE
ROUND THE WORLD.'

LONDON:
JOHN MURRAY, ALBEMARLE STREET.
1859.

5.7   Title page of *On the Origin of the Species by Means of Natural Selection*, London 1859.

The most famous scientific book of modern times was DARWIN's *The Origin of Species* (Fig. 5.7). When it was published in 1859, it caused a tremendous stir, not only in the scientific community, but amongst the general public. The fact was that Darwin had for many years been collecting material for a great treatise on his theory of evolution, but had not published anything when he received from A. R. WALLACE (1823–1913) a brief scholarly paper putting forward exactly the same ideas. This, with a parallel paper by Darwin, was published in the *Journal of the Linnean Society*. He then set to work on this short book, outlining his theory – an abstract of the projected treatise that never did get written. That is perhaps why *The Origin of Species* makes such good reading, and was immediately intelligible to the public. A new scientific theory is seldom stated with such clarity by its original author, and usually takes many years to creep into the public consciousness.

Scientific books continue to be written, published and stored in libraries, although they cost more than can easily be afforded by the individual research worker. Many scientific works are, of course, in the nature of *textbooks*, which expound currently accepted views, within a standard curriculum, for the benefit of students. An undergraduate textbook that has been recommended widely for a popular course in many universities is a valuable literary property, and comes to exert a wide influence. Its point of view will become established as the conventional wisdom in the subject, the original source of a paradigm from which the next generation of research workers will not easily escape. An experienced university teacher talking to schoolmasters, for example, can practically date their undergraduate years by the textbooks from which they evidently drew their scientific opinions!

On the other hand, the scholarly *treatise*, or *monograph* is going out of fashion. The labour of collecting and collating every scrap of knowledge about some scientific topic, presenting perhaps a new integrated viewpoint or a new basic theory, has become too long and arduous. Nobody now waits, like DARWIN, for twenty years before publishing his life's work in a single great book. Perhaps the rate of change of knowledge itself is too great for it to be followed by a single author working on his own. We must make do with *symposia* or other edited compilations of review articles, by many different authors, thrown together and bound up as a book without unity or coherence.

LEARNED JOURNALS

The most important medium of scientific communication is the *primary paper* in a learned *journal*. This was a quite novel invention of the late seventeenth century. We owe more to the great scientific academies for this innovation than for all their other activities (see p. 50). It was, in fact, a natural development, from two types of more private communication. As we have seen, the Secretary of the Royal Society, Henry OLDENBURG, received many letters from Fellows and other scientific correspondents reporting their most recent discoveries. It was natural that he should begin to get these printed and circulated, even though the scientists themselves would be thinking more of presenting their work at much greater length in specially published books. There was also the

Extract of a Letter written to the Editor, from Plymouth, Nov. 2, 1669, by WILLIAM DURSTON, M. D. concerning the Death of the big-breasted Woman, (noticed in N° 52.) together with what was observed in her Body. N° 53, p. 1068.

Elizabeth Travers died on Thursday night, October 21. The next morning I sent for a surgeon, and some others to be present at the opening, and taking off her breasts; though we only took off the largest, which was the left, and having weighed it, we found it 64 pounds weight. Upon opening it, (which we did in several places) we could find neither water, nor cancerous humours, nor any thing vitious, more than the prodigious size; and the tubuli and parenchymous flesh were purely white and solid, and no other than what we see in the soundest breasts of women, or the best udders of other animals. She had lost her appetite and rest several weeks before, and made great complaints of her breasts from their excessive distension, and her whole body was exceedingly emaciated. I have sent you inclosed one measure, which was the breadth of her two breasts (as she was laid out on a table being dead;) I mean, from the further end of the one to the other; which you will find three feet two inches and a half; and another measure showing the dimension of the breasts longwise, viz. near four feet four inches; and a third, giving the dimension of the breadth, viz. three feet four inches and a half.

The right breast we took not off, but guess it might weigh 40 pounds. Some weeks since I began a salivation with her, which lessened her breasts in circumference some inches; but she proving not conformable, I durst not proceed to keep up the flux. But she was wonderfully revived afterwards for some time. She being weary of that course, I caused a caustic to be applied; upon which the eschar fell off, yet nothing issued out of the breast. Then I caused an incision-knife to be used, and made an incision two inches and a half deep (supposing the caustic had not wrought deep enough) but to no more purpose than the former.

An Account of some Books. N° 53, p. 1069.

I. Certain Philosophical Essays, and other Tracts, by the Honourable Robert Boyle, Fellow of the Royal Society. The second edition, enlarged. An. 1669.

This edition is chiefly increased by the addition of a very philosophical discourse on the absolute rest in bodies, wherein the noble author, with his usual modesty and acuteness, delivers his thoughts concerning the intestine motions

Another three-chain structure has also been suggested by Fraser (in the press). In his model the phosphates are on the outside and the bases on the inside, linked together by hydrogen bonds. This structure as described is rather ill-defined, and for this reason we shall not comment on it.

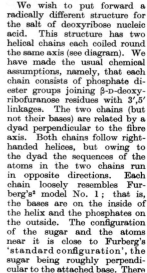

We wish to put forward a radically different structure for the salt of deoxyribose nucleic acid. This structure has two helical chains each coiled round the same axis (see diagram). We have made the usual chemical assumptions, namely, that each chain consists of phosphate diester groups joining β-D-deoxyribofuranose residues with 3′,5′ linkages. The two chains (but not their bases) are related by a dyad perpendicular to the fibre axis. Both chains follow right-handed helices, but owing to the dyad the sequences of the atoms in the two chains run in opposite directions. Each chain loosely resembles Furberg's[2] model No. 1; that is, the bases are on the inside of the helix and the phosphates on the outside. The configuration of the sugar and the atoms near it is close to Furberg's 'standard configuration', the sugar being roughly perpendicular to the attached base. There

This figure is purely diagrammatic. The two ribbons symbolize the two phosphate—sugar chains, and the horizontal rods the pairs of bases holding the chains together. The vertical line marks the fibre axis

5.8   Typical page from early *Philosophical Transactions of the Royal Society.*

5.9   Extract from the paper reporting the discovery of the structure of DNA by J. D. Watson and F. H. C. Crick.

record of meetings – the *Transactions* or *Proceedings* of the Society – which were reported at length, and printed, both as permanent records of research and for the benefit of out-of-town members (Fig. 5.8).

Journals compiled from these various items soon became the standard means of communicating new scientific discoveries. The general format has remained almost unchanged in three centuries (Fig. 5.9). When we think of scientific information we think first of the *primary journals* or *periodicals* that occupy the bulk of any scientific library. These have many significant characteristics, which tell us much about the scientific community and the way it works.

In the first place, a 'paper' is quite short and specific – a few pages of print, summarizing the work of a few weeks or months. Within such a narrow compass, it is impossible, as in a long book, to set up a completely new system of thought from first principles. A typical scientific paper is

King's College, London. One of us (J. D. W.) has been aided by a fellowship from the National Foundation for Infantile Paralysis.

F. H. C. CRICK
Medical Research Council Unit for the
Study of the Molecular Structure of
Cavendish Laboratory, Cambridge.

[1] Pauling, L., and Corey, R. B., *Nature*. **171**, 346 (1953); *Proc. U.S. Nat. Acad. Sci.*, **39**, 84 (1953).
[2] Furberg, S., *Acta Chem. Scand.*, **6**, 634 (1952).
[3] Chargaff, E., for references see Zamenhof, S., Brawerman, G., and Chargaff, E., *Biochim. et Biophys. Acta*, **9**, 402 (1952).
[4] Wyatt. G. R., *J. Gen. Physiol.*, **36**, 201 (1952).
[5] Astbury, W. T., Symp. Soc. Exp. Biol. 1, Nucleic Acid, 66 (Camb. Univ. Press, 1947).
[6] Wilkins, M. H. F., and Randall, J. T., *Biochim. et Biophys. Acta*, **10**, 192 (1953).

5.10 References from the paper, shown in Fig. 5.9.

full of *references* or *citations* to the experiments, calculations, observations, or theories of other people (Fig. 5.10). It does not strike out on its own into the unknown, but timidly takes one little step forward from the base secured by previous research. In other words, modern science is highly *collaborative*, despite all the competition. Everything we do is deeply indebted to, and embedded in, the achievements of our predecessors and contemporaries in our Invisible College. But without good reviews and treatises to integrate all these individual efforts, the progress of knowledge thus tends towards fragmentation and disintegration, so that some subjects became more like miscellanies of fact than organized bodies of knowledge.

On the other hand, a paper can be finished and published in a few weeks or months, so that we get a much quicker turn around in the traffic of ideas. A controversy can boil up and spill over, and be resolved within a year or two, instead of dragging on for half a life-time. Speed of publication is also very important in establishing *priority* of discovery. Much work in the sociology of science is devoted to this topic, which certainly looms large in the pattern of motivation of most scientists. Quick and widespread publication is the surest means of getting *your* ideas out and about, and making sure that other people do not appropriate them.

An important property of the primary scientific literature is that it forms a public *archive* of research. A journal is circulated regularly and automatically, to members of the learned society by which it is published, to private subscribers, and to libraries. It thus makes up a complete series, which is easily indexed by date, volume and page. The

5.11 First page of the reprint of 'Versuche über Pflanzenhybriden' by Mendel, 1866.

conventional format for citations makes free use of such labels: to any physicist the cryptic symbols *Phys. Rev.* **120** (1969), 63 mean a very precise address on page 63, of volume 120 of the *Physical Review*, published in 1969. References to books are much more uncertain and difficult to follow up because there is no guarantee that the particular work will have been purchased by the library that one is searching. Nevertheless, the number of learned journals has become so large that not all relevant scientific papers are noticed by those who ought to know about them. The famous work of MENDEL (p. 33) on the genetics of pea plants (Fig. 5.11) – an apparently obscure problem – was published in aus einer Abkochung sich abscheiden sah, bestand gröstentheils aus *phosphorsaurer Magnesia*, die ich in dieser Wurzel immer in grofser Menge gefunden habe.

Ich werde mich mit einem nähern Studium einiger der oben aufgeführten Stoffe beschäftigen. Ich glaubte, dafs es zweck-mäfsig sey, wenn ich zuvor über ihre Existenz, ihre Darstellung und ihre allgemeinen Eigenschaften Gewifsheit erlange.

Bemerkungen über die Kräfte der unbeleb-ten Natur;
von *J. R. Mayer.*

Der Zweck folgender Zeilen ist, die Beantwortung der Frage zu versuchen, was wir unter „Kräften" zu verstehen ha-ben, und wie sich solche untereinander verhalten. Während mit der Benennung Materie einem Objecte sehr bestimmte Eigen-schaften, als die der Schwere, der Raumerfüllung, zugetheilt wer-den, knüpft sich an die Benennung Kraft vorzugsweise der Be-griff des unbekannten, unerforschlichen, hypothetischen. Ein Ver-such, den Begriff von Kraft ebenso präcis als den von Materie aufzufassen, und damit nur Objecte wirklicher Forschung zu be-zeichnen, dürfte mit den daraus fliefsenden Consequenzen, Freun-den klarer hypothesenfreier Naturanschauung nicht unwillkom-men seyn.

Kräfte sind Ursachen, mithin findet auf dieselbe volle An-wendung der Grundsatz: *causa aequat effectum.* Hat die Ursache *c* die Wirkung *e*, so ist *c = e*; ist *e* wieder die Ursache einer andern Wirkung *f*, so ist *e = f*, u. s. f. *c = e = f ... = c.* In einer Kette von Ursachen und Wirkungen kann, wie aus der Natur einer Gleichung erhellt, nie ein Glied oder ein Theil eines Gliedes zu Null werden. Diese erste Eigenschaft aller Ursachen nennen wir ihre *Unzerstörlichkeit.*

5.12   Mayer's paper on the conservation of Energy.

1866 in the *Transactions of the Brünn Natural History Society* which was not a journal likely to find its way on to the desk of a DARWIN or a HUXLEY. It was not noticed by the scientific world until 1900. The paper of MAYER (1814–78) on the conservation of energy (Fig. 5.12) was published in a chemical journal of 1842 and, being highly speculative in form, was ignored by the physicists, who came to the same conclusions by a different path five or six years later. His failure to win recognition for his work may have caused the mental breakdown that clouded his mind from 1849 until his death in 1878.

Even today, with our *abstract journals, review articles* and *information retrieval systems*, it is important to get one's research published in a reputable scientific journal if it is to be taken notice of. Learned periodicals attempt to maintain a guaranteed standard of scientific quality in the work that is published. A paper in one of the older journals, published by a learned society, would usually bear the name of a member of the society who had 'communicated' it, thus certifying its scientific veracity. Or the editor himself might set his own standard, trusting to his own judgement to exclude trivial work or the irrationalities of cranks. Calling in colleagues to advise him, he thus created the modern system of *referees* – anonymous experts who read papers submitted for publication and recommend acceptance or rejection. The work of MAYER, for example, was rejected by the editor or referees for the *Annalen der Physik* and only accepted by the less appropriate chemical journal. The most vexed and contentious topic in the business of scientific communication is the role of the referees, their danger as censors of new ideas, the procedures for appeal against their decisions, and so on. Only the fact that every referee is himself, on other occasions, an author keeps the system in precarious balance. Yet we cannot dispense with some such system, for it is essential that the material to be found in the 'archival literature' of science should at least seem honest and plausible to those capable of assessing it at the time. The mere fact that an author has a Ph.D. – or is even a distinguished professor – does not ensure that he is free from bias, folly, error, or even mild insanity.

PRODUCTIVITY

Indeed, it is in the personal interest of scientists to maintain a minimum standard of scientific quality in their journals, for this is the material on which their own quality will be judged. A paper conventionally represents a certain amount of successful research by an individual author. The number of papers one has published is thus a crude measure of one's scientific 'productivity'. Let us note, especially, the enormous range of variation of this index (Fig. 5.13). The number of people producing at least n papers decreases rapidly – in fact almost exactly as

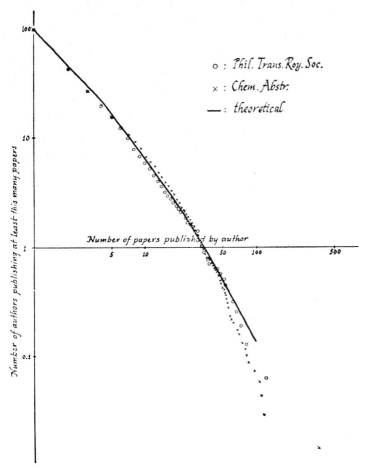

5.13 Number of authors publishing at least *n* papers as a function of *n*.

$1/n^2$, which is an example of a general relationship known as Lotka's Law. In other words, most 'scientists' have published no more than one scientific paper: only about 1% of the scientific community have published more than about ten papers (which would be the least expected of someone of professional standing): only one scientist in a thousand publishes as many as a hundred papers in his lifetime: the record for a single individual is said to be about 1000 publications, which works out at about one paper a fortnight for forty years!

It is obvious, of course, that 'productivity' measured by the mere number of published papers is no absolute index of scientific quality: to quote Derek de Solla Price (*Little Science, Big Science*, p. 40), 'Who dares to balance one paper by Einstein against even a hundred papers by John Doe Ph.D. on the elastic constant of the various timbers (one to a paper)

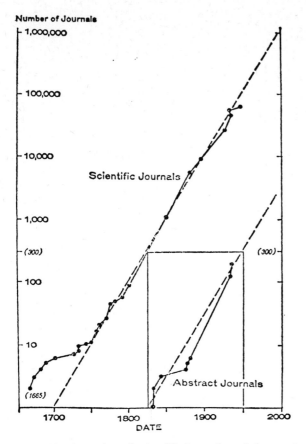

5.14 Total number of scientific journals and abstract journals founded, as a function of date.

of the forests of Lower Basutoland.' But it must be admitted that genuine scientific ability, however estimated, is also distributed in a similar unequal manner. This will be discussed further in the next chapter.

GROWTH AND DIFFERENTIATION

Remember the rate at which science has been growing (Fig. 5.14). This puts a great strain on the system of communication. Every journal must double in bulk every ten years (Fig. 5.15). What happens, of course, is that new journals are started, catering for finer and finer subdivisions of the whole field of knowledge. At the beginning we had the *Philosophical Transactions* covering all branches of learning. Later this divides into physical and biological sections. Then the new learned societies start their own journals, for chemistry, astronomy, geology etc. Nowadays even the major disciplines have been divided into sections: the *Journal of Physics* published by the Institute of Physics in Britain has six sections, which are purchased separately by subscribers according to their specialist interests. In addition there are commercial journals with titles such as the *Journal of Electron Microscopy*, or *Journal of the Physics and Chemistry of Liquids* which provide for particular interdisciplinary Invisible Colleges. This process of growth, differentiation and division is very easy within the present system of primary journals, but leads, of course, to greater fragmentation of the scholarly archives.

There is little doubt that this system of publication of the primary literature of science has great practical advantages. Its long survival, and functional adaptation through such fundamental revolutions as the nineteenth-century academicization and the twentieth-century industrialization of science shows its strength and flexibility. Many people are worried about it, and complain of such disadvantages as the bulk of the journals, the indifferent quality of most papers, the slowness of

5.15   Journal pages published by the American Institute of Physics in 1932 (under one arm) and in 1970 (stacked).

publication etc. But these are mere consequences of the enormous size of science itself and of the modest abilities of most of those who contribute to it. Modern devices such as 'letters' journals or the circulation of preprints in advance of formal publication, do not really avoid the difficulties, and have their own characteristic defects. What is gained in speed is lost in nasty printing, uncritical scholarly standards and impossibility of citation. Various technical 'solutions' to the 'problem of scientific communications' are proposed, such as microfilming, central storage and indexing, computer retrieval etc., but these do not touch the key functions of the attributable, citable, refereed, archival report of research actually done.

The most serious problem, from the point of view of the practising scientist, is to keep up with the literature of his subject, and to be made aware of newly published work that is relevant to his own research. Here again, the *Abstract Journal*, where summaries of all new scientific papers are published under carefully classified headings, is quite an old device that is not easily improved on by computerized retrieval of personalized interest profiles, etc. For those who enjoy designing and selling mechanical gadgets, this is a fertile field, but the real effort is human: careful, thoughtful classification and indexing in the first place, and a little imagination and knowledge of science in searching for what one wants. Much more deliberate attention to the writing of treatises and critical review articles would also help greatly in opening up the archives and finding the nuggets of gold amongst the dross. In any case, this is a technical problem for the scientific community itself, and has little relevance for society at large. It is only a minor symptom of the obesity and flabbiness of science in the contemporary world.

INFORMAL COMMUNICATION

Although scientific knowledge is communicated 'formally' in writing, it spreads widely by means of the spoken word. Not only does the scientist 'write up' his research; he often talks about it, or lectures on it, or discusses it with other scientists in the same field. As we have seen, the original purpose of the first scientific academies was to provide a meeting place for such verbal reports and discussions (Fig. 5.16). Every Thursday afternoon, at 4.30 p.m. after tea, a scientific paper is still read before

5.16 The Royal Society meeting in Somerset House in the early 1800s.

the Royal Society of London, as has been the custom for 300 years. But this is nothing by comparison with the innumerable *seminars* and *colloquia*, given at every university or research institute by speakers from other centres or even other countries. Any scientist of moderate repute is expected to make several such visits each year, at home and abroad. It is a relatively new custom, made easier by air travel and by the lavish financing of modern research, and plays an extremely important part in the rapid diffusion of ideas.

To this one should add the provision of visiting fellowships, exchange schemes and other devices which allow scientists of all grades to spend weeks, months, or years in foreign institutions, learning new ways of thought and teaching their own special skills. Internationalism is implicit in the very definition of science as a body of public knowledge. The members of an Invisible College, colleagues around the world, know each other by actual cooperation in research during such visits; the links are much more personal than merely through the reading of each other's scientific papers. This has always been the case; but the international scientific community, especially in Western Europe and North America, has been greatly strengthened in the past few decades by the continual to-ing and fro-ing of research workers of all nationalities. The most significant generalization one can make about scientific communication is that *ideas move around inside people*.

The most immediate manifestation of this tendency is the proliferation of scientific *conferences* and *summer schools*. In the railway age, the possibilities of rapid travel to bring together all the experts on some particular scientific topic were not adequately exploited. Even in the 1920s and 1930s the *Solvay Conferences*, begun in 1911 (Fig. 5.17) and financed by the great Solvay chemical corporation of Liège, were the only regular international conferences of physicists working on such fundamental problems as nuclear structure and statistical mechanics. After the Second World War, however, the whole picture changed. With funds from various sources – government research organizations, private industry, foundations, international organizations – the calendar became more and more crowded with meetings on every conceivable scientific topic. The format varies, from highly specialized and exclusive

5.17 The Solvay Conference 1911/12. Showing from left to right: standing, Goldschmidt, Planck, Rubens, Sommerfeld, Lindemann, De Broglie, Knudsen, Hasenohrl, Hostelet, Herzen, Jeans, Rutherford, Kamerlingh Onnes, Einstein, Langevin; seated, Nernst, Brillouin, Solvay, Lorentz, Warburg, Perrin, Wien, Madame Curie, Poincaré.

5.18 First General Conference of the European Physical Society. Taken during
V. F. Weisskopf's plenary lecture at the Palazzo Vecchio, Florence.

conferences of 50 or 100 high-level experts discussing deeply in an hotel in
the Alps (with long walks in the afternoons, to preserve the atmosphere
of informality) to congresses of 5000 miscellaneous crystallographers or
biochemists gathering for interminable parallel sessions in a dozen
lecture rooms scattered about a great city such as Vienna or Tokyo. A
senior scientist of international standing can spend almost his whole life
attending such meetings, skipping back and forth round the world in and
out of aircraft and hotels, giving his standard lecture, answering the
standard questions, and dining well with a group of colleagues from
other countries following the same circuit (Fig. 5.18). Or you may find
him at a summer school, lecturing daily for a week or two whilst enjoying
a vacation in Florida, or Sicily, or Norway, or Bangalore. The younger
men go, and listen, and meet their contemporaries, and begin to feel that
they too belong to the international community of science.

It is easy to poke fun at the whole business of scientific conferences.
There are obviously too many of them: the programmes are too
crowded; the discussion is seldom very elevated. On the one hand they
are taken too seriously, as if they were really important; on the other
hand there is an element of frivolity in supposing that anything useful can
be said in a ten-minute talk or learnt in a fortnight of lectures. They have
become a part of the trappings of modern science – an excuse for the
conspicuous travel that now replaces conspicuous expenditure as a symbol
of power and success in worldly affairs.

Yet they serve a vital function in the transfer of knowledge, binding
the international scientific community together by ties of personal

friendship and mutual understanding. Not to be able to attend the international conferences in one's subject, not to be able to meet one's scientific contemporaries around the world, is to be condemned to isolation, to provincialism, and eventually to the frustration of all one's efforts to keep up with the moving frontiers of research. This is the plight of so many scientists in developing countries (see chapter 11), who have so little money for travel, and who have so far to go. It is also, to a considerable extent, the plight of science in the Soviet Union and its satellites, where political obstacles to travel have serious consequences for the actual conduct of research, beyond the obvious harm they do to the morale of the research workers themselves.

POPULARIZATION OF SCIENCE

I have been discussing communications *inside* the scientific community. What about bringing scientific knowledge to the general public. This is not a trivial task – nor is it a new one. 'It is not easy to devise a cure for such a state of things [the declining taste for science]; but the most obvious remedy is to provide the educated classes with a series of works on popular and practical sciences, freed from mathematical symbols and technical terms, written in simple and perspicuous language, and illustrated by facts and experiments which are level to the capacity of ordinary minds.' That comes from the *Quarterly Review* for February 1831. Or go back another century to SWIFT, for whom Gulliver speaks:

The sum of his discourse was to this effect: That about forty years ago, certain persons went up to Laputa, either upon business or diversion, and after five months' continuance, came back with a very little smattering in mathematics, but full of volatile spirits acquired in that airy region. That these persons, upon their return, began to dislike the management of everything below, and fell into schemes of putting all arts, sciences, languages, and mechanics, upon a new foot. To this end, they procured a royal patent for erecting an academy of projectors in Lagado; and the humour prevailed so strongly among the people, that there is not a town of any consequence in the kingdom without such an academy. In these colleges, the professors contrive new rules and methods of agriculture and building, and new instruments and tools for all trades and manufactures; whereby, as they undertake, one man shall do the work of ten; a palace may be built in a week, of materials so durable as to last for ever without repairing; all the fruits of the earth shall come to maturity at whatever season we think fit to choose, and increase an hundred-fold more than they do at present; with innumerable other happy proposals. The only inconvenience is, that none of these projects are yet brought to perfection; and, in the meantime, the whole country lies miserably waste, the houses in ruins, and the people without food or clothes. By all which, instead of being discouraged, they are fifty times more violently bent upon prosecuting their schemes, driven equally on by hope and despair: that, as for himself, being not of an enterprising spirit, he was content to go on in the old forms, to live in the houses his ancestors had built, and act as they did, in every part of life, without innovation. That some few other persons of quality and gentry had done the same, but were looked on with an eye of contempt and ill-will, as enemies to art, ignorant, and ill commonwealth's men, preferring their own ease and sloth before the general improvement of their country.

His lordship added, that he would not, by any further particulars, prevent the pleasure I should certainly take in viewing the grand academy, whither he was resolved I should go. He only desired me to observe a ruined building, upon the side of a mountain about three miles distant, of which he gave me this account: That he had a very convenient mill within half a mile of his house, turned by a current from a large river, and sufficient for his own family, as well as a great number of his tenants. That about seven years ago, a club of those projectors came to him with proposals to destroy this mill, and build another on the side of that mountain, on the long ridge whereof a long canal must be cut, for a repository of water, to be conveyed up by pipes and engines to supply the mill: because the wind and air upon a height agitated the water, and thereby made it fitter for motion; and because the water, descending down a declivity, would turn the mill with half the current of a river whose course is more upon a level. He said that, being then not very well with the court, and pressed by many of his friends, he complied with the proposal; and after employing an hundred men for two years the work miscarried, the projectors went off, laying the blame entirely upon him, railing at him ever since, and putting others upon the same experiment, with equal assurance of success, as well as equal disappointment...

...we came back to town; and his excellency, considering the bad character he had in the academy, would not go with me himself, but recommended me to a friend of his, to bear me company thither. My lord was pleased to represent me as a great admirer of projects, and a person of much curiosity and easy belief; which, indeed, was not without truth; for I had myself been a sort of a projector in my younger days.

This academy is not an entire single building but a continuation of several houses on both sides of a street, which, growing waste, was purchased and applied to that use.

I was received very kindly by the warden, and went for many days to the academy. Every room hath in it one or more projectors; and, I believe, I could not be in fewer than five hundred rooms.

The first man I saw was of a meagre aspect, with sooty hands and face, his hair and beard long, ragged, and singed in several places. His clothes, shirt and skin were all of the same colour. He had been eight years upon a project for extracting sunbeams out of cucumbers, which were to be put into vials hermetically sealed, and let out to warm the air in raw inclement summers. He told me he did not doubt, in eight years more, he should be able to supply the governor's gardens with sunshine at a reasonable rate; but he complained that his stock was low, and entreated me to give him something as an encouragement to ingenuity, especially since this had been a very dear season for cucumbers. I made him a small present, for my lord had furnished me with money on purpose, because he knew their practice of begging from all who go to see them.

5.19 Conduction of sound by solid bodies.

It is evident from this passage that the Fellows of the Royal Society had singularly failed to convey to the Dean the inner significance of their research and the benefit that it would assuredly bring to the nation. If such a well-educated man showed such ignorance of science as to ridicule it in this manner, what must have been the state of the populace at large.

Books of popular science are not a new invention. This one (Fig. 5.19) dates from the late nineteenth century, and is full of delightful pictures and conjuring tricks. The main contribution of our own era to this sort of literature is colour printing.

Another nineteenth-century speciality was the popular scientific lecture. The *Royal Institution* was founded in 1799 by Count RUMFORD (p. 74) as a sort of technical college, where science of a socially relevant kind could be brought to the people. In fact it became a small but very successful research laboratory where DAVY and FARADAY (p. 57) made their great discoveries. A masterly style of popular lecturing was developed there (Fig. 5.20), illustrated with numerous elegant demonstrations, and soon became immensely fashionable amongst the educated classes of Victorian London (Fig. 5.21). This tradition continues to our

5.20 Experiments at the Royal Institution, 1802. Gillray, 'Scientific Researches! – New discoveries in PNEUMATICS! – or – an Experimental Lecture on the Powers of Air – '

5.21 Michael Faraday lecturing at the Royal Institution.

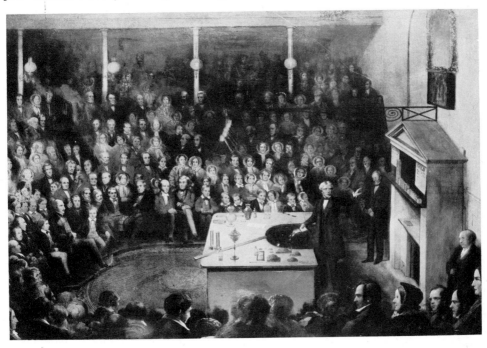

own day. Television broadcasts of the Royal Institution Christmas lectures for children are amongst the best popular science programmes now being produced in Britain.

THE BRITISH ASSOCIATION

Another interesting development was the *British Association for the Advancement of Science*, which was founded in 1831, but reached its height of power and fame in the later part of the century. As its title suggests, it had social relevance as its central aim, and included many lay members as well as professional scientists and academics. The special recipe of this unusual organization was to meet each year for a week in some provincial city, where numerous lectures would be given by experts on all sorts of topics (Fig 5.22). The idea was to bring science and society together for the benefit of the nation: in fact, its aims were very close indeed to the ideal of social responsibility in science. But the tendency was for the professionals to take it over and to treat it as a general meeting place for the scientific community, where new discoveries could be reported and general discussions arranged on controversial questions.

5.22   British Association in 1865. (See *Nature*, **246** (1973) 388, for possible identity of figures.)

5.23 Samuel Wilberforce, Bishop of Oxford. Caricature by 'Ape' (Carlo Pellegrini) from *Vanity Fair*.

For British science as a whole it was the equivalent of a conglomeration of annual conferences, where each Section could independently indulge in its specialist mysteries.

The most famous occasion of the 'British Ass.' was at its Oxford meeting in 1860 when the Bishop of Oxford, Samuel WILBERFORCE ('Soapy Sam') (Fig. 5.23), bitterly opposed to DARWIN's theory of the evolution of the species, asked sarcastically of T. H. HUXLEY (see p. 137) whether he traced his own descent from the apes through his father or his mother. HUXLEY answered that 'if he had to choose as an ancestor either a miserable ape or an educated man who could introduce such a remark into a serious scientific discussion, he would choose the ape'.

The decline of the 'BA' in recent years is a serious loss to science and to society. It is essential that the scientific community as a whole should have a general forum for its views, which are not adequately represented by the Royal Society on its lofty pinnacle nor by the various more specialized learned societies. Unfortunately, the peripatetic annual meeting is no longer a satisfactory form of assembly, being ponderous, too demanding of time, and slow to react to external challenges. If the BA did not exist then it would have to be invented – but in a more efficient and up-to-date form, as a bridge between professional science and the community at large. The American Association for the Advancement of Science, with a similar constitution, seems to be adapting more smoothly to this important role.

SCIENTIFIC JOURNALISM

Perhaps this is only a symptom of a general trend. The task and duty of explaining science is slipping out of the hands of the research workers themselves into the care of a body of professional middle men – science journalists, writers of popular books and TV producers. Scientists of high standing are now usually too specialized in their interests, too competitively involved, too taken up with administration, travel and other business, to put the effort into the interpretation of their discoveries to the general public. So now we find 'communication experts' who themselves specialize in writing newspaper columns and constructing TV programmes about science. We also have specialized journals – in particular *Scientific American* – concerned solely with the exposition of recent scientific work in non-specialist language. The technical skills employed in these media are very high indeed: the best programmes and articles are beautifully illustrated and of high scientific precision. What is not clear is the extent to which they are watched or read by non-scientists. The registered circulation of *Scientific American* is about half a million copies a month: would not such a number be taken up by the world-wide community of professional scientists, engineers, and other technically skilled people, together with students in universities, technical colleges and high schools? It turns out to be a full-time job just to keep up with the research literature in one's own field: high class 'popular science' is invaluable as background reading for the working scientist himself, even though it may be somewhat beyond the man in the street.

THE PROBLEM OF COMMUNICATION

The communication of modern science to the ordinary citizen, necessary, important, desirable as it is, cannot be considered an easy task. The prime obstacle is lack of education. It is almost impossible to talk about the most elementary new discoveries to people who have not grasped the rudiments of the older knowledge of the subject. School science is not very easy to teach, and seldom gets over the first barriers of ignorance to form a coherent system. The language in which most modern scientific ideas are expressed and grasped takes years to learn, and cannot be paraphrased for easier comprehension. Even science students are abysmally ignorant of disciplines other than their own. Physics students can reach an honours degree without any notion at all of the function of DNA in reproduction: no doubt students of zoology are equally hazy about the nature of radio waves or the electron theory of chemical valence. Until we provide a thorough course of general science in all secondary schools, outlining the main features of the world of nature as now known, with some attention to the historical, social and technological significance of this knowledge, it is a waste of effort to put on marvellous programmes on pulsars, or Continental Drift, or the structure of the brain, for a general audience.

There is also the difficulty of making scientific discoveries interesting and exciting without completely degrading them intellectually. There are very real issues of incompatibility of temperament between the research worker and the journalist. There is a real contradiction between sensationalism and scientific caution, between colourful exaggeration and sceptical precision, between the modesty and impersonality considered appropriate to the scientist and the cult of flamboyant personality of the popular press. It is a weakness of modern science that the scientist shrinks from this sort of publicity, and thus gives an impression of arrogant mystagoguery; but one must understand his genuine revulsion against the style of person and of writing that he must affect in the world of the mass media. It is essential to recognize the depth and width of this psychological gulf if one is to bridge it successfully.

6 AUTHORITY AND INFLUENCE

And so in matters animal, vegetable and mineral I am the very model of a modern
Major-General.   *W. S. Gilbert*

INEQUALITY IN THE REPUBLIC OF SCIENCE

In principle the scientific community is a democratic republic. Anyone
with new ideas or with valid criticism of current ideas can publish his
work. Brilliant scientific advances have been made by young men of great
originality. In 1830 Évariste GALOIS (Fig. 6.1), a student in his first year
at the École Normale, was killed in a duel at the age of 21. The night
before, he wrote a letter explaining his theory of the solution of algebraic
equations. This was the foundation of an entirely new branch of pure
mathematics, the theory of groups, which has completely transformed
our thinking about the nature of algebra and geometry.

Such early brilliance does not always go unrecognized. B. D.
JOSEPHSON (1941–   ) (Fig. 6.2) was still an undergraduate at Cambridge when he pointed out that an important experiment using the
Mössbauer effect to test the theory of relativity was subject to a
significant temperature effect – a point that had been overlooked by the
experienced physicists who had planned the experiment. He went on to
discover a fundamental new quantum phenomenon in superconductors
whilst still a research student, was elected to the Royal Society in 1970, at
the age of 29, and shared the 1973 Nobel Prize for Physics.

In practice, however, the scientific world is as highly structured as any
other human group; a few scientists are very much more equal than
others. To understand science in relation to society, we must understand
the role of the leaders, or, as the term goes, the scientific 'authorities'.

We have already noted the great range of individual scientific
achievement. Although the quantitative measurement of the number of
published papers (Fig. 6.3) is a spurious index of scientific ability in any

6.1   Évariste Galois aged 16.   6.2   Dr B. D. Josephson, FRS, Nobel laureate.

6.3   Number of authors publishing at least *n* papers as a function of *n*.

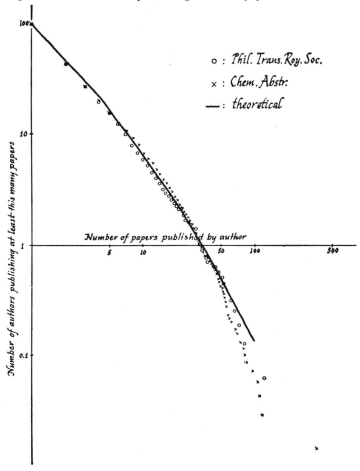

particular case, a range of 1000:1 in achievement is as natural in science as it is in any other creative human activity. V. GINZBURG, in his obituary notice of L. D. LANDAU (1908–68), reported that LANDAU classified physicists on a logarithmic scale.

This means that a physicist, say, of the second class has accomplished (precisely *accomplished*, we are dealing only with accomplishments) a tenth as much as a first class physicist. On this scale, EINSTEIN was of class one half, and BOHR, SCHRÖDINGER, HEISENBERG, DIRAC and a few others first class. LANDAU placed himself in a two and a half (i.e. only one hundredth of an EINSTEIN!) and only some ten years ago, satisfied with some of his work (I recall this conversation, but forget the work involved) he stated that he worked his way up to second class.

Quality measurements of this kind are difficult and dangerous. We all know of discoveries that are highly acclaimed in their own day which have turned out to be sterile or even false. Many of the most famous discoveries were not original or unique to those who get the credit for them. We have already noticed (p. 103) the tragedy of MAYER whose discovery of the principle of the conservation of energy was overlooked when first published. In fact this principle was quite clearly stated by *twelve* different scientists, independently of one another, between 1830 and 1850, before it became common knowledge.

On the other hand, some really first class work (on the LANDAU scale) received little public recognition at the time. For example, neither Willard GIBBS (1839–1903) nor Ludwig BOLTZMANN (1844–1906), the sources of almost everything worth knowing about statistical thermodynamics, received the Nobel Prize. And let us remember Alfred WEGENER (p. 78) who was treated as a crank all his life for his advocacy of the hypothesis of Continental Drift.

For these reasons, a highly developed system of prizes, rewards and public honours for scientific work is dangerous and dishonest. To make 'stars' of a few, and thus, by implication, to degrade those who work hard with less good fortune is not healthy. Enrico FERMI (p. 84) is made to look a fool in the plumed hat of a Royal Academician (Fig. 6.4): the 'Road to Stockholm' for the Nobel Prize is not open to many of those who truly deserve such an honour. The whole business of pecking orders and quality grading in science, as in art, is altogether too uncertain to be justifiable. It would be better to preserve in public the fiction of equality of standing at the professional academic level and not to give additional

6.4 Fermi. Members of the Royal Academy of Italy. Fermi on far right.

social emphasis to the natural differences of ability which are all too evident to those inside the system.

INTELLECTUAL AUTHORITY

Nevertheless, scientific authority is very real, and comes in many different forms. Consider first the authority of intellectual power, of which the supreme example is Isaac NEWTON.

As we have seen (p. 43) NEWTON lived a very quiet life as a Cambridge don until he was 54. He was not exactly a recluse, but he communicated to the scientific world mainly through his letters and books. He had many correspondents but no personal pupils, and he founded no 'school' of research. In 1696 he became Master of the Mint, and carried through with great efficiency a reform of the coinage, but for this he would scarcely deserve more than a passing reference in history. As a civil servant, for example, he cannot be compared with Christopher WREN (1632–1723) who rebuilt London after the Fire, nor with Samuel PEPYS (1633–1703) who reconstructed the Royal Navy. For many years he presided over the Royal Society, but more as a landmark or monument than as a creative administrator.

Yet his influence was enormous. In science itself his discoveries and theories revolutionized astronomy, physics and mathematics. The development of these subjects during the next 150 years seems no more than the working out of his ideas. The great French mathematical physicists of the late eighteenth century, LAPLACE and LAGRANGE, regarded their own numerous discoveries merely as necessary consequences of the 'Newtonian synthesis' of classical mechanics.

Remember, however, that in his great work on optics, NEWTON finally came down in favour of the particle theory – as if light were like a stream of bullets. This, unfortunately, was incorrect since it could not explain diffraction phenomena; NEWTON's enormous authority probably retarded the introduction of the wave theory of light, which was not accepted until about 1820.

As the greatest scientist of his era, NEWTON became a symbol of *English* science, and a special source of national pride. The present-day publicity game of counting Nobel prizes as if they were Olympic medals or wins at football is not new. For many political leaders, the cultivation of science by the state is justified as a means of encouraging and rewarding the big names in this international competition. National pride can sometimes interfere with scientific judgement. NEWTON's mathematical symbolism was retained in England for several centuries, even though it is not really as good as LEIBNIZ's method of indicating the operations of the calculus. LEIBNIZ, after all, was a foreigner (p. 44) and not British.

But NEWTON's intellectual authority had a far wider influence. His name is usually invoked as the prime cause of the change in 'man's view of nature' that took place in the seventeenth and eighteenth centuries. NEWTON's success in explaining many details of the motion of the earth, the moon, the planets and the sun made it seem plausible that the whole universe could be described completely by mathematics and that it all ran 'like a piece of clockwork' with very little intervention by even a benevolent Deity. This was not by any means what NEWTON himself believed, nor what he said. He had slightly odd theological views, but did not publish any of his vast researches into Biblical chronology. He made much more modest claims in his scientific work, which was, of course, quite incomprehensible to almost everybody except a skilled mathematician. Yet a garbled and popularized version of this work became a ruling principle of European thought in the 'Age of Reason'. It served as well for the agnostic who wished to push God completely out of the picture as for the theologian who could point to the extraordinary efficiency of His construction of the Universe. It became an excuse for treating men as machines and formed a basis for new doctrines about space and time.

The effect of analytical mechanics on the metaphysical foundations of European civilization is just as important as its technological consequences. The name of NEWTON became a symbol of the intellectual authority of science, which could now claim parity with religion as a prime mover in human society. Characteristically, this authority has been personified and focussed through its most famous individual creator, whose scientific discoveries and theories are generalized, oversimplified, and served up as dogmas. Although domineering and unscrupulous in defence of his scientific priorities in his old age, NEWTON cannot be blamed for the many illegitimate uses for which his authority was borrowed; but leadership in science carries a personal responsibility to state clearly the limits of application of our current understanding.

The modern equivalent of NEWTON is, of course, Albert EINSTEIN (1879–1955) (Fig. 6.5). He was an unworldly, gentle, saintly person, who

6.5   Einstein riding a bicycle at a friend's home near Los Angeles, February 1933.

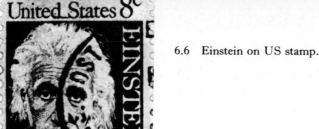

6.6 Einstein on US stamp.

founded no school of disciples and held no powerful public office. Modern mass communications made him a world-famous figure (Fig. 6.6), the scientific equivalent of Picasso in art or of Stravinsky in music. His great head of white hair, his quiet countenance, the simplicity of his manner, dress and way of life, all contributed to the perfect image of the scientific genius, living in another plane of existence from mere mortals. Say '*scientist*'; you will get the perfectly conditioned response '*Einstein*'.

EINSTEIN's authority was expressed through his superb achievements in theoretical physics. These add up to a small number of papers and a few short books. To the general public these are quite incomprehensible: even amongst professional physicists his later work would still be considered very stiff going, and not entirely successful. The simplified versions of his theory of relativity are no more than superficially plausible: a long education in physics and mathematics is essential if one is to appreciate the beauty, originality and inward power of his ideas.

Yet this work is regarded with almost superstitious awe by the public. The mystical formula $E = MC^2$ has become symbolic of the power of man over nature, and of science over mankind. The transformation of this hypothetical relationship between mass and energy into the terrible reality of the atomic bomb was the traumatic event of modern science and of modern society. EINSTEIN himself was too evidently a sweet and benevolent person to be blamed as an evil genius, but his elevation into the patron saint of physics may be due as much to fear as to respect.

In the spheres of philosophy, politics and even aesthetics, EINSTEIN's name is invoked to justify all sorts of bogus principles of 'relativism' which have nothing to do with the very particular sense in which the dimensions of space and time are 'relative' to the motion of a physical observer. But a mysterious theory which seems to upset all our conventional notions of the structure of the everyday world was just what the doctor ordered for those who proclaim new fashions in the *zeitgeist*. After the cold clockwork of 'Newtonianism' and the struggle for survival of 'Darwinism', how very convenient to have an excuse for social and artistic revolution in 'Einsteinism'. The question is not the extent to which this scientific theory has had any influence on the general philosophy of modern society; it is the degree to which ideological movements with other social sources now tend to claim legitimacy by association with scientific ideas. COPERNICUS, NEWTON and DARWIN personify successive stages in the take-over by science of many aspects of nature that were formerly controlled by theology. EINSTEIN does not seem to represent a forward – or backward – step along that particular path.

These remarks are abstract and schematic: but EINSTEIN's personal story is also a very instructive example of another aspect of the relationship between the scientific genius and society. The racialism of Nazi Germany from 1920 onwards was expressed very forcibly against EINSTEIN as the leading Jewish intellectual of his epoch. Violent antisemitism turned upside down the patriotic pride of the German people in their scientific pre-eminence. Deliberate attempts were made to prove that relativity was false precisely because it had been discovered by a Jew, who was accused of 'contaminating Aryan physics' and so on. EINSTEIN was not, himself, much interested in politics – he was perhaps more of a Swiss at heart than a German – but in 1933 he was driven into exile in the United States. He thus escaped more lightly from the Hitler terror than the majority of European Jews; but the episode demonstrates, in its pathological extremes, the extent to which science is caught in the net of cultural nationalism and racialism. The scientist ignores the nature of the society in which he lives at his own peril.

EINSTEIN's only public office, outside the academic world, was his first job on graduating from the Technical University at Zürich: not having made sufficient impression on his professors to be appointed a scientific assistant, he took a post in the Swiss Patent Office, where he had adequate spare time to pursue his scientific research. He had no temporal power. He was never a manager, a research director, a member of parliament, or a government minister. Yet he could not escape the problems and responsibilities of social action. In 1939 he was asked by

F.D. Roosevelt,
President of the United States,
White House
Washington, D.C.

Sir:

Some recent work by E.Fermi and L. Szilard, which has been communicated to me in manuscript, leads me to expect that the element uranium may be turned into a new and important source of energy in the immediate future. Certain aspects of the situation which has arisen seem
to call for watchfulness and, if necessary, quick action on the part
of the Administration. I believe therefore that it is my duty to bring
to your attention the following facts and recommendations:

In the course of the last four months it has been made probable -
through the work of Joliot in France as well as Fermi and Szilard in
America - that it may become possible to set up a nuclear chain reaction
in a large mass of uranium,by which vast amounts of power and large quantities of new radium-like elements would be generated. Now it appears
almost certain that this could be achieved in the immediate future.

This new phenomenon would also lead to the construction of bombs,
and it is conceivable - though much less certain - that extremely powerful bombs of a new type may thus be constructed. A single bomb of this
type, carried by boat and exploded in a port, might very well destroy
the whole port together with some of the surrounding territory. However,

I understand that Germany has actually stopped the sale of uranium
from the Czechoslovakian mines which she has taken over. That she should
have taken such early action might perhaps be understood on the ground
that the son of the German Under-Secretary of State, von Weizsäcker, is
attached to the Kaiser-Wilhelm-Institut in Berlin where some of the
American work on uranium is now being repeated.

Yours very truly,

A. Einstein

(Albert Einstein)

6.7   Letter from Einstein to Roosevelt, 2 August 1939.

SZILARD (1898–1964) and WIGNER (1902–), two other distinguished European physicists in exile, to sign a letter (Fig. 6.7) addressed to President Roosevelt drawing the attention of the United States government to the danger of the Nazis constructing a uranium bomb. EINSTEIN was not, in fact, deeply interested or committed politically, but he knew that he could not avoid this action, by which his scientific authority set great social forces in motion. From this letter came the American atomic bomb project, and all that followed. In 1945 EINSTEIN wrote to Roosevelt – again at SZILARD's suggestion – urging him to prevent the bomb from being used against Japan: but this letter was found unopened on Roosevelt's desk the day he died. Just before his own death, in 1954, he agreed to sign the manifesto organized by Bertrand RUSSELL (1872–1970), which was the origin of the Pugwash Scientific Conference (see p. 335). By these actions, and by his moderate support for Zionism, he showed his liberal, undoctrinaire response to the tragic circumstances of his times; the most refined and pure scientist of his day, a man made primarily for private scholarly concentration of mind, could no longer stand aloof; as the leader of one of the prime forces in modern society, he had to exert his influence for the good as he saw it.

The most profound influence of science on social thought came, of course, from the biological revolution of Charles DARWIN. The actual occasion of the publication of *The Origin of Species* has already been described (p. 98). The theory of the evolution of species by natural selection was so clearly and simply explained in language that was perfectly intelligible to any educated man, that it was possible, almost, for everyone to make up his own mind on the subject. It is interesting to note that DARWIN, although immensely respected, admired and honoured for the remaining twenty years of his life, lived in quiet retirement (Fig. 6.8) and was not turned into a demigod. The controversy concerning the theory of evolution rumbled on for several generations until it came to be accepted as the simple truth. We teach it now to children in primary schools, long before they can learn the significance of Newtonian analytical mechanics. 'Darwinism' is so much more directly relevant to man in society than any philosophy based on physics that we can scarcely imagine ourselves nowadays thinking otherwise.

But remember that this theory, too, is concerned with very particular biological phenomena, in a well-defined scientific context. DARWIN said little about the evolution of human institutions, or of cultural traits, or of 'racial' characteristics within the human species. Yet his 'authority' has been used over the past century to justify a variety of malicious or ridiculous doctrines in the wider field (see p. 283). The authenticity of evolution by natural selection for biological systems makes it all too easy to slip over into a plausible argument for the application of the same mechanisms in society at large. Consciously or unconsciously, our modern civilization uses such arguments to justify many of its activities and attitudes.

It is also interesting to recall that DARWIN owed a great deal to the famous book by Thomas MALTHUS (1766–1834) – the *Essay on Population*, published in 1798 – which drew attention to the enormous potential rate of reproduction of a human population that was not limited by its food supply. MALTHUS was a somewhat pessimistic political writer, much out of favour with revolutionaries and 'progressives'. Some modern commentators have even tried to set a black mark against DARWIN for drawing on such a tainted source. But the relation between social philosophy and scientific theories is not quite so simple as that.

6.8 Darwin's study at Down House, Downe, Kent. Engraving by A. H. Haig.

6.9 Justus von Liebig.

6.10 Liebig's laboratory at Giessen, 1842.

DARWIN's theory of *biological* evolution stands on its own feet, on the biological evidence and by its own direct logic. MALTHUS provided the germ of an idea, but that does not mean that DARWIN was accepted, or was attempting to justify, his gloomy prognosis of the future of man. This is the sort of pitfall that is often encountered in the game of searching deeply for social and economic influences in the history of science.

AUTHORITY AS A TEACHER

But scientific authority takes other forms. Consider, for example, Justus von LIEBIG (1803–73) (Fig 6.9). He was one of the greatest chemists of the nineteenth century with many original contributions to organic chemistry and biochemistry. But his special place in the history of science stems from the 'school' of chemical research that he founded in 1824 at the university of Giessen where he was professor for twenty-eight years. This was the first laboratory for the *teaching* of science (Fig. 6.10). LIEBIG's students matured to become professors in other universities where they reproduced his laboratory under their own direction. From their pupils, in turn, grew the great German chemical industry of the late nineteenth century, to which we have already referred. As we can see

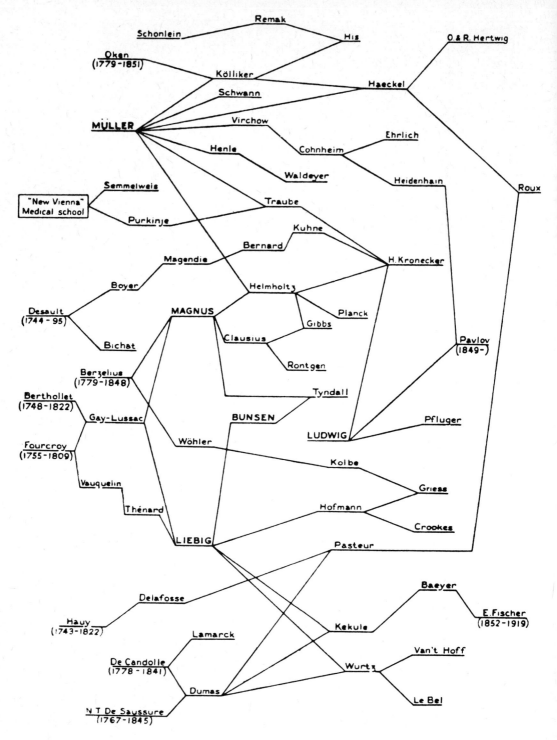

6.11 'The connection of master and pupil', mainly nineteenth century.

from this chart (Fig. 6.11), showing the connection of master and pupil in that century, LIEBIG was the central figure on the chemical side. He taught almost all the very best chemists of the next generation, and was a scientific 'grandfather' to many more.

This is a characteristic form of authority within the scientific community. LIEBIG was, indeed, recognized in the world outside the university – he was made a Baron for his pains – but his greatest influence was on the intellectual shape of the science itself, not so much by revolutionary discoveries but as a brilliant teacher. It is from such men that we acquire our 'paradigms' – the basic principles, techniques, and concepts on which a given science is currently believed to depend. The very power and efficiency of their teaching, the clarity of exposition that they achieve in their lectures, the gentle patronage that they extend to their pupils, are irresistible. The pupils of such a master learn their subject as it is seen through his eyes – often without all the doubts and uncertainties that the teacher himself may personally feel. It is interesting to note, by the way, that LIEBIG believed in 'spontaneous' chemical fermentation of yeast, wine etc., and was long a bitter opponent of PASTEUR, who eventually showed that these phenomena were produced by micro-organisms (p. 62). Consulting the chart, we see that PASTEUR belonged to a different 'school' of chemistry; it was thus not surprising that he held heretical views.

It has often been observed that famous scholars have famous pupils. In our own day we may study the 'genealogy' of Nobel Prize winners (Fig. 6.12) and notice the tendency for this premier award to run in 'families'. What is the explanation of this phenomenon? A significant factor, obviously, is that a distinguished scientist, lured into a post at a famous university, has the pick of the best young men of his day as pupils, and is thus likely to have taught some of the winners in the next generation. One may also remark, cynically, that the great men must use their power to favour their own pupils. But there is a genuine factor that is often underrated by those unfamiliar with scholarly pursuits. Original, creative research is a very subtle art, that does not come naturally to most people. It requires more than intelligence, imagination and persistence. One has to learn to be self-critical, sceptical and yet self-confident in the pursuit of worthwhile ideas. From a great master, these qualities can be learnt by imitation. To make an absolutely first-rate contribution to

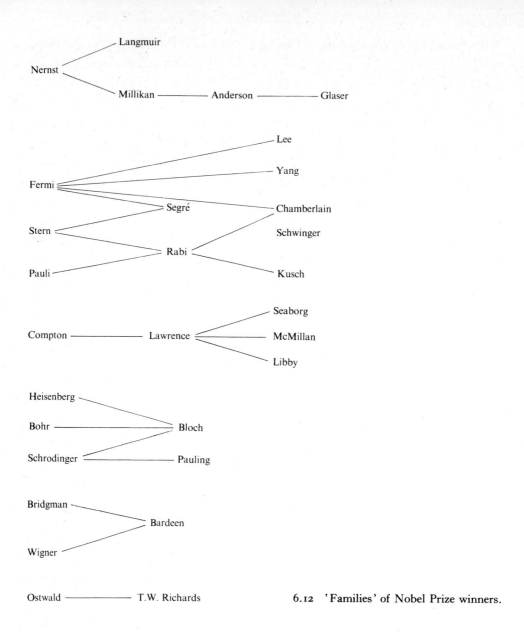

6.12 'Families' of Nobel Prize winners.

science, one must, for example, have a subtle grasp of what constitutes a truly significant question, and what is needed to give an adequate answer. It could be that this is the quality that really singles out the Nobel Laureate, and this is the inheritance that he passes on to his pupils.

At a lower level, it is abundantly clear that scientific research as a craft is learnt by apprenticeship, and that it takes many years to acquire sufficient skill to practice this craft on one's own. The transference of scientific authority from generation to generation and from one country to another is not achieved merely by setting up schools of study and research. It demands a supply of fully trained and experienced scientists who have had the time and the opportunity to come to maturity themselves before attempting to pass on their expertise to others. This is a problem to which we shall return in chapter 11 when we discuss the growth and spread of science into the 'Third World'.

ADMINISTRATIVE AUTHORITY

As well as intellectual authority and great teachers, the scientific community has its administrative leaders. Remember Sir Joseph BANKS (p. 54) (Fig. 6.13) who was President of the Royal Society from 1778 (he was then 35) until his death in 1820. He was a very rich man, with a passion for botany, who had contributed in money and in his own person to the scientific success of Cook's expedition to the South Seas (Fig. 6.14). For forty-two years he ran the Royal Society, benevolently,

6.13   Sir Joseph Banks. Portrait by Thomas Phillips.

6.14   'The great South Sea Caterpillar, transformed into a Bath Butterfly.' Gillray 1795.

6.15 Frontispiece to 'Peter's Prophecy...or, an important Epistle to Sir J. Banks, on the approaching election of a president of the Royal Society'. 1788.

6.16 G. Cruikshank (after Captain Marryat), 'Landing the Treasures'.

generously, with good-humoured paternalism. At his own expense he entertained scientific guests at weekly breakfasts and evening 'Conversations' (Fig. 6.15). He was the dictator of English science and the acknowledged expert on all branches of practical knowledge, and he took a leading part in the foundation of the great Botanic Gardens at Kew. He encouraged the colonization of Australia and the transplantation of useful plants from one region to another of the world (Fig. 6.16). BANKS was a tough, masterful character in the President's chair, and played a strong game against his enemies: on one occasion he managed to dismiss the Council of the Society, just as they were planning to dismiss him. But he also made good use of his money in many acts of philanthropy on behalf of poor scientists such as William SMITH (p. 53).

In any human community, a man with eminent gifts of management and sociability is likely to gain a great deal of power. He is so useful: he gets things done; he oils the works; he wins friends; he influences people. The scientific community needs such people as much as any human group. But observe that he becomes the channel through which the world of science communicates with the power of the state. Too much urbanity and worldly wisdom, in the absence of scientific integrity and insight, could be disastrous in such a key position.

AUTHORITY AS PUNDIT

On the other hand, Thomas Henry HUXLEY (1825–95) (Fig. 6.17) was a self-made man. Like BANKS and DARWIN, he won scientific fame for his biological observations on a voyage to Australia and became Professor of Natural History at the Royal School of Mines. But the publication of *The Origin of Species* (p. 117) swept him into the world of theological debate, popular science, education and journalism. In the 1870s and 1880s HUXLEY was one of the best-known scientists in England – not with the pure intellectual authority of DARWIN but as the energetic public man, member of a dozen Royal Commissions, a key figure in the creation of a state system of elementary education, writer on religion, on science etc. He had a fluent pen, liberal principles and a strong, clear intellect. In a sarcastic mood we may refer to him as the Public Relations Officer of Darwinism and of Victorian science: it is more accurate to see HUXLEY

6.17   Caricature of T. H. Huxley by 'Ape' (Carlo Pellegrini), *Vanity Fair*, 1871. 'A great Med'cine-Man among the Inqui-ring Redskins.'

6.18   Photograph of Lord Kelvin with his compass.

as the prototype of the 'socially responsible' scientist. His sincere but rather self-conscious religious radicalism mirrors the naive political radicalism of some of our own distinguished contemporaries.

AUTHORITY AS TECHNICAL INNOVATOR

Victorian England is a good habitat for scientific authority. Consider, as a final example, William THOMSON, First Baron KELVIN (1824–1907) (Fig. 6.18) who was Professor of Natural Philosophy at the University of Glasgow for fifty-three years. In the world of pure science, KELVIN is known as a first-rate mathematical physicist – not quite the equal, to my mind, of James CLERK MAXWELL or of Lord RAYLEIGH (1842–1919), but certainly amongst the 'immortals'. But perhaps this ability alone would not have qualified him for the singular preferment of a Barony – the first British scientific Lord. What was recognized by the society of his day were his practical researches in electricity, magnetism and mechanics.

His theory showed how to build a workable transatlantic telegraph. He designed instruments for the measurement of electric power supply. He completely redesigned the ship's compass according to advanced physical principles. He invented a sounding apparatus (Fig. 6.19), tide gauges etc., and studied the mathematical theory of the wake of a ship. What could be more unconsciously, perfectly, Marxian than the paper he read before the British Association in 1881 showing that 'the most economical cross section area for a conductor for power transmission is that for which the cost of energy losses in a given time is equal to the interest on, and depreciation of, the capital involved, in that time'? KELVIN is the prototype of the pure scientist as technological consultant, contributing, with all his intellectual brilliance, personal charm, and modesty, to the political–military–industrial complex of his day. He has usually been regarded, quite rightly, as a considerable benefactor to mankind, who put his scientific authority to direct human use.

6.19   Kelvin's sounding machine in use, from *Admiralty Manual of Seamanship*, 1942.

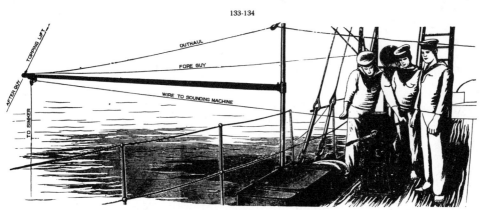

SOUNDING FROM THE FORE BRIDGE—WIRE RUNNING OUT.

The scientific authorities whom we have encountered up to now were all great scientists. But F. A. LINDEMANN (1886–1957) (Fig. 6.20) owed his power in the scientific world to other qualities. A wealthy Englishman (his father came from Alsace), he did research in physics at Berlin, and made quite a name for himself as a young man of considerable scientific promise. During the First World War he showed courage and ability in experimenting with the flight characteristics of aircraft at the Royal Aircraft Factory at Farnborough. In 1921 at the age of 33 he became Professor of Physics at Oxford, where he was expected to transform the old Clarendon Laboratory into a first-rate research institution, competitive with the Cavendish Laboratory at Cambridge. In this he succeeded – but much more by the hospitality offered to several outstanding German refugee scientists after 1933 than by his own personal contributions.

Much more significant was LINDEMANN's close friendship with Winston Churchill. Just before the Second World War he became involved in the scientific problems of Air Defence. The story of his conflict with Henry TIZARD (1885–1959) (p. 312) over radar is one of those juicy bits of gossip that everyone remembers. Whatever the rights and wrongs of it, when Churchill became First Lord of the Admiralty in 1939, and Prime Minister in 1940, he took LINDEMANN along as his personal scientific adviser. In this position he was the most powerful scientist in Britain, controlling the general direction of war research, the deployment of scientific manpower and resources, the detailed development of strategic weapons, and even having considerable personal responsibility for the interpretation of intelligence reports and tactical appreciations. It was a new kind of post; and because this war was fought so heavily with scientifically designed weapons – radar, rockets, tanks, submarines, aircraft etc. – it became of enormous importance. The question arose, for example, whether it was more effective to fight an aggressive war of mass bombing of Germany or to put the effort into defence against U-boats. In the end it was probably LINDEMANN's statistical calculations, based upon incomplete intelligence reports, that threw the weight on the bombing policy, which was later shown to have

6.20   Lord Cherwell seated next to Winston Churchill, on a visit to Harwell in 1948.

been more costly and less militarily effective than he argued. War is not so amenable to rational design as experimental physics.

Nevertheless, the outstanding achievements of British scientists during the Second World War owe much to LINDEMANN's forceful, tough-minded leadership. He was not perhaps the leader they would themselves have chosen, for his reputation in academic research had fallen very low, and he had made no impression as a technological innovator or as a shrewd administrator. But he was a genuine physicist and, armed with political and bureaucratic authority on a level with the military chiefs of staff and cabinet ministers, he represented the scientific community in the central offices of government power, and made sure that its human and intellectual resources were fully used.

After the War LINDEMANN was made Lord CHERWELL, and served for a while as a Conservative Cabinet Minister. He was an odd personality in many ways, and the dust of controversy has not settled around his place in history. In a later chapter we shall consider in more detail the relationship between the scientist and war. We shall also meet quite a number of other characters whose authority in the world of science does not rest on their eminence in research but has its roots in the bureaucratic power conferred on them by the organs of the state. But LINDEMANN's career illustrates the simple fact that the scientific community is a body of citizens who must eventually subordinate themselves to the corporate will of society at large. The significant feature was that this force was channelled through a man of acknowledged scientific competence and integrity, rather than through a stratum of laymen or through a bunch of self-serving nonentities, as can happen in more corrupt state machines.

J. Robert OPPENHEIMER (1904–67) was another rich youth who showed great scientific promise when he went to Göttingen in the 1920s and worked with Max BORN (1882–1970) on the newly discovered quantum theory. He returned to the United States in 1929 and became the leader of theoretical research in nuclear physics at the University of California at Berkeley. He acquired a very high reputation as a man who understood deeply the fundamental mathematical problems of this rapidly developing subject, and, although his own contributions to physics did not compare in originality with those of his German contemporaries, he built up a first-rate graduate school from which came many of the best American theoretical physicists of the next generation.

But he did not remain an austere professor of an esoteric subject. In 1942 he was commissioned by the US Government to direct a newly created laboratory at Los Alamos, in the desert wilderness of New Mexico. This laboratory was charged with the sole task of designing and constructing an atomic bomb. The story of the discovery of nuclear fission, and of the secret moves in Germany, Britain and the United States towards the use of this natural phenomenon in a weapon of war, is too long and involved – and too important historically – to be summarized here. Suffice it to say that OPPENHEIMER was charged with one of the central tasks in the whole project – and that he accomplished it extraordinarily well. He gathered together a brilliant team of academic scientists, American and European, and somehow got them working cooperatively together. Physicists of the Nobel Laureate grade –

6.21 Robert Oppenheimer and General Leslie Groves in the Alamogordo desert.

LANDAU's Second Class, shall we say (p. 122) – do not take kindly to the bureaucratic ways of the US Army, under the command of General Groves (seen here with OPPENHEIMER on the site of the first test explosion of a nuclear bomb (Fig. 6.21)). OPPENHEIMER created an intellectual and moral atmosphere in which they could give of their best and get the job done. Yet he was a pure academic scientist without experience of engineering or administration. Nor was he by nature an open 'clubbable' man, at ease with other people; he did not always conceal an intellectual arrogance that often made bitter enemies.

After the War, as Director of the Institute for Advanced Study at Princeton, he could well have retired with honour into academic life. But he had become too deeply involved with political and administrative power, and became chief scientific adviser to the US Atomic Energy Commission, responsible for the further development of nuclear weapons and of more constructive uses of atomic energy. The decision to proceed with the construction of thermonuclear weapons – i.e. the Hydrogen Bomb – was taken in an atmosphere of court politics and high intrigue, with all the forces of militarism, nationalism, Cold War treachery and scientific responsibility locked in conflict. This, again, is an historical episode of such significance that one must study it for oneself. For OPPENHEIMER himself, it brought deep tragedy, for in 1954 he was accused of being a security risk and dismissed from his position as consultant adviser to the AEC.

The 'trial' of J. Robert OPPENHEIMER is one of the most important historical events of our era. It is almost certain that he was entirely loyal to his country, in war and in peace. He had had communist friends before the War, but was not really a man of conventional political interests. He scarcely read a newspaper and saw himself in the image of the entirely dedicated pure scientist, like NEWTON or EINSTEIN. He behaved foolishly about some rather petty security investigations – but all too obviously out of arrogance rather than with deliberate intent to deceive. The charges were trumped up, as a manoeuvre to get rid of him, in the unscrupulous manner of political life at the top. His partial rehabilitation in 1963 is reasonable evidence on that account. In fact it was nothing more than a dirty, rotten bit of political trickery, activated by malice and vindictiveness, but without much influence in the actual course of events, for he was not, in fact, standing in the way of some great change of policy, and could have been eased out or bypassed without all the drama.

It is highly significant, however, as an illustration of the deep problems of the modern relationship between the scientific community and the organization of the state. OPPENHEIMER was by no means the sole or chief representative of 'science' in a conflict with 'government'. He had, in fact, become part of 'government' itself, and his opponents were as much scientists, like Edward TELLER (1908–   ), as 'politicians'. Nor was it a question of the conscience of the humane scholar against militaristic barbarism. OPPENHEIMER had had to search his heart very deeply when he realized what had been done by dropping an atomic bomb on Hiroshima, but he had certainly not withdrawn into a pacifist frame of thought. The fact that the whole scientific community did not rise up in his support is also significant. He was by no means the hero of all the physicists, who were deeply divided themselves on the issue.

John VON NEUMANN (1903–57) once said: 'In modern science, the era of the primitive church is over, and the age of the Bishops is upon us'.

OPPENHEIMER is like one of those characters whom Gibbon describes with such delicious irony in the *Decline and Fall of the Roman Empire*. He is no scientific martyr, like some of those who suffered under Hitler or Stalin, but a bishop, saintly or otherwise as you wish, defeated in a party brawl over doctrine. He reminds us that scientific authority, as it acquires secular power, can lead to corruption and its own degradation. The incorporation of science into the state, the formal acknowledgement of the scientific community as an estate of the realm, inevitably generates these tragic contests for power amongst the leaders of this estate.

7 FROM CRAFT TO SCIENCE

Every man desires to gain wealth that he may give it to the doctors, the destroyers of life; therefore they ought to be rich.   *Leonardo da Vinci*

TECHNIQUES AND TECHNOLOGIES

Let us now follow another thread through the labyrinth. Practical crafts or *techniques* are characteristic of all human societies. Specialized skills have existed, and continue to exist, independently of a formal body of precise knowledge or theory, and are passed on by demonstration, practice, or personal experience from generation to generation. Typical modern examples are household cookery, sewing, gardening, fly-fishing or fox-hunting.

But the development of a complex civilized way of life is characterized by the appearance of persons who specialize in the performance of particular skilled activities – i.e. of *experts*. This elementary consequence of the division of labour enhances technical progress by competition and by the pooling of knowledge. New techniques arise, demanding more deliberate training and formal education of each succeeding generation. The accumulation and consolidation of practical knowledge creates a new level of expertise, embodied in the high-order profession of the *teacher* of the craft. Amongst the teachers we observe an interest in fundamental principles and in explanations for the success of the practice that they profess. This leads eventually to a 'scientific' appproach to technical problems.

This is the true meaning of *technology* – strictly speaking, the 'science' of a craft, art, or technique. The word is now used very loosely to denote the actual practice of the skill thus acquired, just as the word 'science' itself is commonly applied to any self-consciously rational or rationalized activity. As we have seen, the improvement of techniques and of technologies is intertwined with the growth of 'pure' science; in this chapter, however, we consider the development of 'scientific technologies' from 'practical crafts' as an historical and social phenomenon in its own right, with its own typical stages and problems.

MEDICINE

Our first example is a major human craft – medicine. All human societies, from the most primitive, have techniques for dealing with sickness or bodily injury (Fig. 7.1). To our way of thinking these do not always seem very helpful, but they are usually based upon rational religious notions about evil spirits or have intelligible roots in sympathetic magic. The professional 'medicine man', with special techniques and apparatus, is still to be found today in many parts of the world (Fig. 7.2) though he is often scarcely distinguishable from the harmful purveyor of

7.1 Cures for madness amongst the Indians of Paria. From: *Cérémonies et coutumes de tous les peuples du monde représentées par des figures dessinées de la main de Barnard Picart.* Amsterdam 1723–43.

the power of magic. The skills required to practise these arts are not learnt from books or from a college curriculum, but are secret 'mysteries' passed on from father to son, or to an 'adopted son', or through a long private apprenticeship. Each practitioner is thus also cast in the role of a teacher, within the family circle, just as a good wife and mother teaches her daughters to sew and to cook.

The most famous early teacher of medicine with many pupils was reputed to be HIPPOCRATES (Fig. 7.3), who lived on the Greek island of Cos around 400 BC. But he may not really have been an actual single person. The name has become attached to a considerable body of writings, probably by many different authors over a wide period, systematically expounding the practice of medicine as a rational art. This work is remarkable for its observational accuracy and objectivity: the scientific spirit within the culture of ancient Greece is as much in evidence in the 'Hippocratic Collection' as in the books of ARISTOTLE or of EUCLID. In clinical medicine, for example, it mainly recommended the physician not to interfere with the 'healing power of Nature', and to take advantage of local conditions of climate.

7.2 African doctor and peddler of magic remedies.

7.3 Hippocrates, from a Byzantine manuscript, *c.* 1342.

This writing is best known to us for the code of professional conduct – the so-called *Hippocratic Oath*: Everybody is familiar with noble sentiments such as

whatsoever I shall see or hear in the course of my profession, as well as outside my profession in my intercourse with men, if it be what should not be published abroad, I will never divulge, holding such things to be holy secrets.

But notice also the conventional guild or trade union declaration:

To hold my teacher in this art equal to my own parents; to make him partner in my livelihood; when he is in need of money, to share mine with him; to consider his family as my own brother, and to teach them this art, if they want to learn it, without fee or indenture; to impart precept, oral instruction, and all other instructions to my own sons, the sons of my teacher, and to pupils who have taken the physician's oath, but to nobody else.

A body of professional experts is always tempted to exploit its monopoly of the secrets of its technique. Greek medicine had, however, reached such a sophisticated level of skill that the healing power of the

7.4 Galen seated among physicians and botanists talking to Dioskourides. Dioskourides, *De materia medica.*

physician lay in his long apprenticeship and personal experience rather than in any particular secret formula which could be looked up in a book. The fact that these writings became *public* knowledge was not a serious threat to the monopoly position of the medical guild. But this was, of course, a decisive step forward in the development of scientific medicine, for it encouraged the publication of new observations, new techniques and new theories. We take it for granted that new technical knowledge in the field of medicine will be published and freely shared as soon as possible, and that it should not be kept as a trade secret, or sold for cash, or licensed for use like a patent right. The position is by no means the same in other useful arts, where it is considered perfectly proper for the individual practitioner to profit from any private information he may gain to his own advantage.

From the Hippocratic tradition sprang the extensive and competent medical technique of the Roman Empire. For those who could afford their services, the doctors of the Graeco-Roman culture of the Mediterranean in the first few centuries AD provided as effective treatment as was available in modern Europe until about the end of the eighteenth century. Whatever their practice, however, they were divided into many conflicting schools of theory. There were *Dogmatists, Empirics, Methodists, Pneumatists* – and *Eclectics* who claimed to combine the views and virtues of all the others. Thus, the 'Pneumatists' believed all disease to be due to a disturbance of the 'airy spirit' or 'pneuma' that was inhaled with the breath and permeated the organism. This disturbance could be detected by feeling the pulse. The 'Empirics' believed, more modestly, that the doctor should rely on experience alone, labelling the disease and the corresponding remedy, thus reducing medicine to a rule of thumb.

The most famous medical figure of that time was GALEN (AD 131–200) (Fig. 7.4) who was born in Pergamos in Asia Minor. He became the leading practitioner of his day amongst the Roman aristocracy, and did not hesitate to attack the various medical sects with his brilliant pen. But GALEN was also an outstanding investigator, who described with great accuracy and completeness what he actually saw in his dissections of animals, which he incorporated into his description of the human body. He thus founded the science of *anatomy*. He also constructed an ingenious system of *physiology*, dominated mainly by pneumatic concepts but used to explain many new and correct observations that he had himself made.

GALEN's extraordinary position as the ultimate authority in medicine for the next 1300 years did not depend merely on his brilliant researches and plausible theories. His writings were so forceful, dogmatic and apparently infallible that they were taken to be the literal truth, despite many errors that could have easily been checked at any time. There has never been a more extreme example of a scientific or technological paradigm, personified in a great teacher and then mummified in the written word. Contrary to what one might imagine, the 'expert' is often the very last person to be forced by his direct practical experience to doubt the theoretical principles on which he was trained. In this respect technology is not, in fact, more empirical, down to earth, sceptical and practical than highbrow pure science.

From Hellenistic medicine we move naturally to its heir and successor, the medicine of the Arab world. In the works of AVICENNA (980–1037), the great physician, philosopher and encyclopaedist of Bokhara, we observe this at its apex. Note the clear technical diagrams in this edition

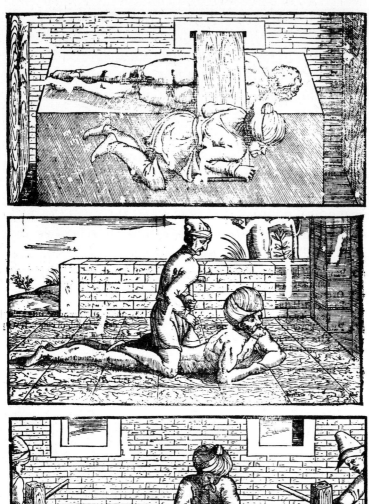

7.5 Dislocation treatment. From: Avicenna, *Canon of Medicine*, Latin ed. 1608.

(Fig. 7.5). A physician could use these to real effect. The general scheme is Galenical in theory, but there had obviously been a lot of quiet research and invention to improve practical techniques in surgery. A considerable body of anatomical and physiological knowledge had accumulated and was being taught systematically. In this era, great *hospitals* were founded at the major Arab cities of Baghdad, Damascus, Cordova and Cairo, where the teaching of medicine went on along with the care of the sick. These would constitute veritable *schools*, with a more permanent institutional character than a single practitioner with a few apprentices or pupils.

The first real medical school in Christian Europe grew up at Salerno, some time in the eleventh century, under the enlightened rule of the Norman kings of Sicily. The Italian coast around Naples has been considered an agreeable region since Roman times and, as a health resort, Salerno attracted many wealthy people with their attendant private physicians. It was a natural setting for the teaching of medical practice. Through Sicily there also came some of the first translations from Arabic into Latin of the Greek and Arabic medical literature, which was, of course, far in advance of the European medicine (Fig. 7.6) of that day.

The School of Salerno flourished for only about two centuries, but it is important as the prototype of the modern university. The medical course lasted for five years with an additional year of practice under supervision,

7.6 Dislocation treatment (reduction) by traction and pressure on the spine, from a fifteenth-century manuscript.

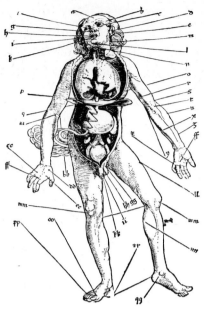

7.7 Bloodletting chart. Woodcut by Johannes Wechtlin, from a treatise on surgery by Guy de Chauliac, inserted in Hans von Gersdorff, *Feldtbuch der Wundartzney*, Strassburg 1540.

and was followed by an examination granting the degree of Doctor of Medicine. The great mediaeval universities founded at Bologna, Paris, Oxford etc. in the twelfth and thirteenth centuries copied this pattern when they incorporated medicine as one of the major faculties of a *studium generale*.

This formalization and systematization of a craft was not necessarily progressive. Some forms of ignorance, incompetence or fraud could be prevented, but where the doctrine itself was at fault it was now made much more rigid by deliberate careful teaching. The teacher of medicine himself, elevated to the exalted rank of lecturer or professor, was no longer necessarily in active practice. He was liable to spend more of his time preparing his lecture notes out of ancient treatises (Fig. 7.7) or theorizing further within the orthodox Galenical framework of thought, than in observing patients, doing experiments or making anatomical studies. Public dissections were rare (Fig. 7.8): clinical research did not exist: book learning, even in this most practical of human arts, was enthroned. It is noteworthy that the main advances in medicine during the Middle Ages came in surgery, where the consequences of faulty technique could not fail to be observed.

7.8 Anatomy lesson of Mundinus at Padua, from Ketham, *Fasciculus Medicinae*. Venice 1500.

7.9 Military surgeon extracting an arrowhead. Woodcut by Johannes Wechtlin, from Hans von Gersdorff, *Feldtbuch der Wundartzney*, Strassburg 1540.

One of the greatest surgeons that ever lived was Ambroise PARÉ (1510–90) who had no formal medical education, but learnt his craft as a barber's apprentice, in the great hospital at Paris, and as a military surgeon (Fig. 7.9) in the Italian campaign of 1536–45. He took surgery almost as far as it could go without anaesthetics and antiseptics. His classic work was on the treatment of gunshot wounds – an example of the stimulus of war on technology.

Modern medicine really begins in the year 1543, with the publication of VESALIUS' marvellous treatise on anatomy (p. 96). It is true that dissections and anatomical drawings of great realism and accuracy were being made about 1500 by such artists as DÜRER, MICHELANGELO, RAPHAEL, and, above all, LEONARDO DA VINCI (p. 69); but these were adjuncts to the art of painting and made no impression on the doctors.

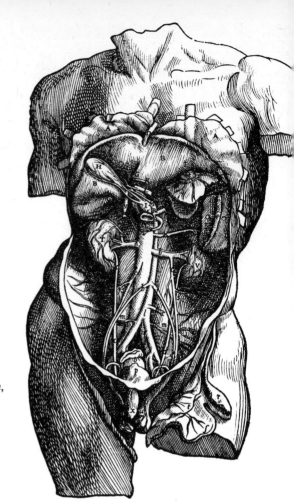

7.10   Viscera figure from Vesalius' *Fabrica*,
Book v. Basel 1543.

VESALIUS studied medicine at Paris, where he was subjected to lectures
drawn from the works of GALEN, describing anatomical structures that
the Professor himself could not point to in the human body. VESALIUS
then taught anatomy himself at the University of Padua, and, with the
aid of several excellent artists, at the age of 29, he produced *De humanis
corporis fabrica*, where, for the first time in a published work, men could
see the actual structure of their own bodies (Fig. 7.10). The importance
of this work lies not only in its great beauty and observational accuracy
but also in that it struck deliberately to the very heart of dogmatism in
medical science. VESALIUS challenged the orthodox professors to refute
him by public demonstration and openly confounded them. Their
failure to be immediately converted, and their righteous indignation at
being put in the wrong, discouraged him from further anatomical
studies; but the battle had in fact been won. From that time onward, the

science of human anatomy was built entirely on accurate observation and not by reference to learned opinion or ancient authority.

Nevertheless, the hold of GALEN over *physiology* was not broken until the work of William HARVEY (1578–1657), the physician to James I and Charles I. After study of all the relevant published material, very detailed anatomical studies on dead and living animals, and many ingenious experiments, in 1628 he wrote his masterpiece on the circulation of the blood (Fig. 7.11). In a mere 52 pages he demolished the doctrines of his day (according to which the blood ebbed and flowed in the veins, whilst the arteries carried the 'vital spirits'), and set out the essentials of our modern view on this fundamental physiological process. Here, at last, was a positive achievement of the new science of direct observation and experiment that had been proclaimed by Francis BACON. But notice again that HARVEY was a practising physician; although he gave regular lectures on medical subjects, he was not committed to the teaching of the elementary principles of medicine to a class of uncritical students.

A College of Physicians was founded in 1518 in London by Henry VIII, with power to license practitioners, examine prescriptions etc.

7.11 Harvey, diagram of the circulation of the blood. From: *De Motu Cordis*..., 1628.

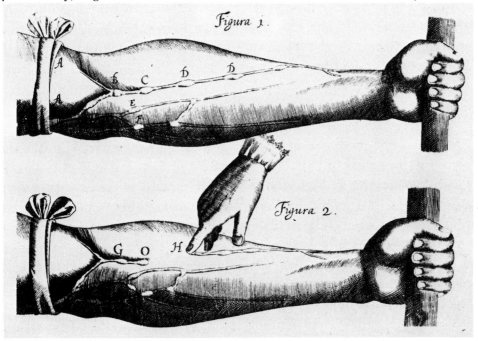

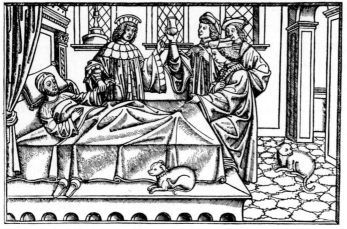

7.12 The Consultation (Italian School, sixteenth century). Woodcut from title page of Panthaleo's *Pillularium*. Pavia 1516.

(Fig. 7.12). This is another characteristic stage of 'professionalism' – the creation of a monopoly for accredited experts. We now take for granted that this training will take place in public institutions, such as universities and teaching hospitals, which are themselves inspected and licensed by some national authority. It is well to remember, however, that 'private' medical schools existed in England until the nineteenth century, and that the standards of medical education and practice in the United States were of deplorable variability until well into the present century.

HOW SCIENTIFIC IS MEDICINE?

From the eighteenth century, medical training has remained much the same in spirit (Fig. 7.13). It is founded on anatomy and physiology, which are treated as logically consistent natural sciences with a firm experimental basis. This is followed by clinical studies, under the direct guidance of experienced practitioners, where the emphasis is on practical lore and on empirically successful techniques (Fig. 7.14). In every era it is properly emphasized that medicine cannot 'yet' be reduced to a complete science that covers all cases in theory, and that the skilled practitioner must be ready to use his personal initiative and intuition to deal with the actual world. Clinical research in medicine also has a strong

7.13   Medical students at an anatomy lesson.

7.14   The College of Physicians. Coloured etching by Rowlandson & Pugin, 1808.

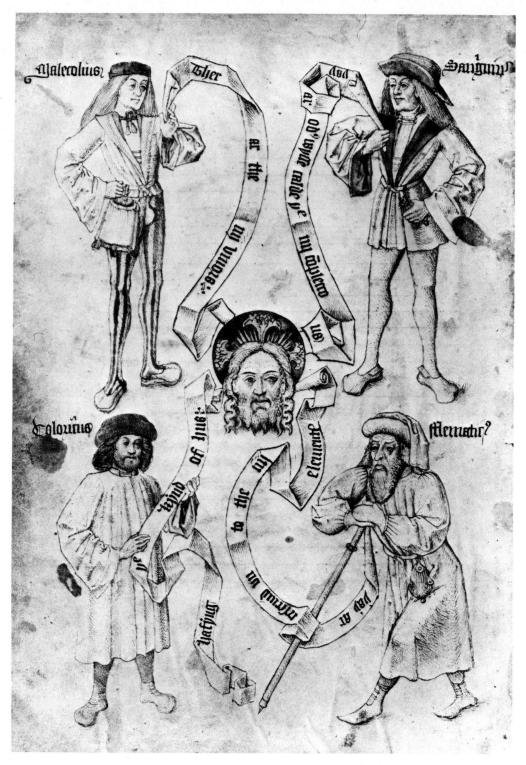

7.15   The Four Temperaments. From *The Guild Book of the Barber Surgeons of York, c.* 1500.

empirical tradition. Successful treatments discovered by accident or by inspired guesswork are just as valuable, in themselves, as those arrived at by rational deduction from theory. The purpose of medical education and research is to cure people of disease, not merely to add to our knowledge of human biology.

But this pragmatic attitude has not preserved medicine from ill-founded dogmatism (Fig. 7.15). Experiment can be dangerous when it might jeopardize the safety of the patient. The doctor who advises a 'sound' treatment based upon conventional principles need not fear to carry the responsibility for failure. In any practical craft where decisions *must* be taken daily on inadequate evidence, there is a perfectly proper tendency towards theoretical conservatism. To bolster up the confidence of the doctor himself (let alone that of his patient!) it is more comfortable to lean upon some general theory that appears to justify the treatment than to fall into scepticism and moral insecurity.

Consider, for example, the plight of the eighteenth-century physician. He had a choice of medical 'systems' on which to base his diagnosis and treatment. Following BOERHAAVE (1668–1738) he might believe in '*solidism*', whereby the human body is held to be composed of solids immersed in humours. Disease was caused by, say, a change of humidity in the air, which upset the equilibrium of the humours (Fig. 7.16). To

7.16   Uroscopy and pulse. Caricature from Metz Pontifical. French manuscript of 1316.

restore the equilibrium, the physician should prescribe sedatives, tonics etc.

Or he might adhere to HOFFMANN (1660–1742) who held that the human body is a machine whose movement is governed by the flow of humours. Disease was mainly due to disturbance of this flow at given points on the digestive tract, which would affect the nerve fluid etc. HOFFMANN was a great advocate of hot and cold baths.

CULLEN (1712–90) on the other hand believed that pathogenic causes acted directly on the nervous system. This was a theory derived from the physiological concept of the 'irritability' of the 'fibres' that were supposed to be the essential constituents of all living organisms. Quinine had an important role in his treatments to reduce these reactions.

BROWN (1735–88) arrived at quite different conclusions from the same general principles. He considered irritability an essential part of life and would prescribe stimulants such as alcohol or electric shock to maintain it. But one of his former disciples, RUSH (1766–1837), reversed all of BROWN's conclusions, although he would often arrive at the same practical treatment.

The system of BROUSSAIS (1772–1838) was based upon the effect of external heat on the humours. To cure a disease it was necessary to reduce the local heat by copious blood letting, with applications of numerous leeches to the head and stomach (Fig. 7.17).

It is easy to laugh at these fantastic notions. It is wiser to recall the thousands of patients who were undoubtedly harmed by absurd treatments based upon the various theories without regard to common sense. Everybody knew perfectly well that loss of blood as a result of a wound could be a serious danger to life – yet we find the doctors, right into the nineteenth century, advocating extensive blood letting even in the treatment of pneumonia. As Dr Carlo Cippola has pointed out, it was not until about 1850 that DIETL in Vienna and BENNETT in Edinburgh showed by simple inspection of hospital statistics that the doctors were killing twice as many patients by their treatments as would have died naturally from this disease. It is humbler to suppose that many of the 'cures' for our psychological or social ills, proposed by our own experts upon the basis of equally half-baked general theories, are of no greater merit.

The most dangerous situation arises when the experts are themselves firmly bound together in a craft, guild, or trade union. So long as the conflicting schools quarrel and argue openly, the layman can pick and choose between them or at least maintain a healthy scepticism. But when a particular sect gains control of the professional institutions, and appoints itself the guardian of 'sound practice', the general public should beware. The history of medicine is full of such episodes. Let us recall, for example, the opposition of orthodox medicine to the new psychiatric methods of Sigmund FREUD (1856–1939). This opposition went considerably beyond mere intellectual criticism and forced the psychoanalysts to set themselves up as an independent body of practitioners outside the medical profession. The cry of 'no quackery' has often been used by complacent conservatives to suppress radical technical innovation.

The history of medicine is instructive because it shows the extraordinary difficulty of putting a practical craft upon a sound scientific basis. Almost all improvements in medical practice until very recent times were achieved by direct observation, by experiment, or by simple deduction from a wide variety of familiar facts.

It was not until 1865 that a radical change of medical technique was

7.17 A surgeon bleeding a lady, by Abraham Bosse.

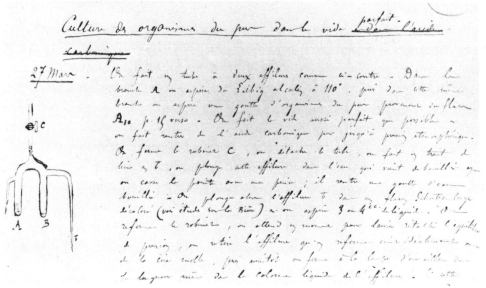

7.18 Pasteur's notebook: Cahier d'expérimentations (February–May 1878) p. 50, 'Culture des organismes du pus dans le vide avec un tube à double effilure'.

derived from a fundamental biological principle. In that year Joseph LISTER (1827–1912), Professor of Surgery at Glasgow, took note of PASTEUR's evidence in favour of the germ theory of disease (Fig. 7.18), and began to experiment deliberately with techniques of antiseptic surgery (Fig. 7.19). When the cure for cancer finally comes, it will be interesting to see whether it arises naturally out of clinical medicine as a brilliant 'invention' or whether it will be based theoretically upon some new fundamental mechanism of molecular biology.

The relationship between the organized medical profession and the general community is also of significance. As an indispensable expert, the doctor has power which he has not hesitated to use to his economic and social advantage. The craft monopoly of the Royal College of Physicians in Tudor times did the doctors no harm; modern successors, such as the American Medical Association, use all the tactics of the political game to preserve the privileges and incomes of their members. We may rightly celebrate the foundation of a professional society as the 'coming of age' of a practical art – for example, the formation of the various Engineering Institutions in the middle of the nineteenth century represent a new stage of precise, analytical 'scientific' technique in an ancient craft. But there is a real danger when the political functions of such an organization as a *trade union* are confounded with its role as a *learned society* concerned with the advancement of expert technique. As a trade union, unity of purpose and a hierarchical structure of power on a democratic base may be essential: as a learned society it needs to be an open forum, where a multiplicity and diversity of opinions may be freely expressed. There is good sense in keeping these functions quite separate. Looking at the history of the scientific technologies, we may well ask whether it is wise, as some politically minded scientists would insist, to turn all the learned societies in 'academic' science into professional associations lobbying for rights of tenure, pay scales, and against the war in Vietnam.

METALLOGRAPHY

Medicine, you may say, is peculiarly difficult as a theorizing science because of the complexity of its subject matter. Let us turn, therefore, to a very simple physical system produced by a very ancient craft. The fabrication of objects out of iron and steel goes back more than 3000 years,

7.19   Lister, operation scene *c.* 1880, showing use of carbolic spray.

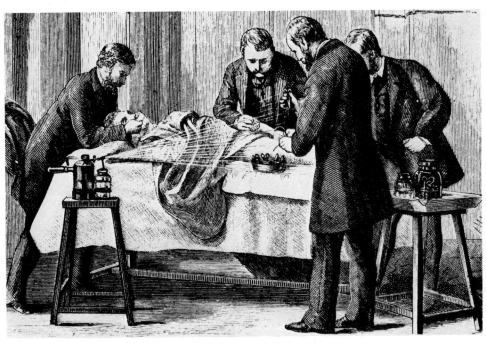

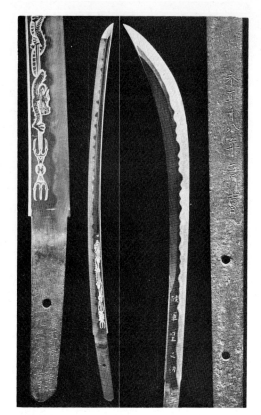

7.20 Japanese swords.

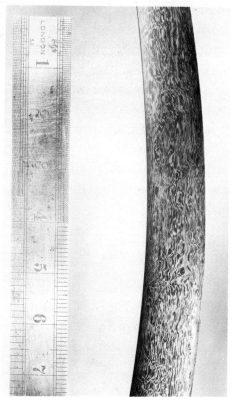

7.21 Seventeenth-century Persian sword, showing damask pattern.

and was extremely important for human society, in peace as in war. Great skills were acquired by the smiths, by accumulated experience and bold experiment. A fine sword was the product of extreme artistry and concentrated effort. These mediaeval Japanese swords (Fig. 7.20) are unsurpassed, not only for their strength and sharpness, but for the beautiful ornamental finish that brings out the 'grain' of the steel. Remember the fame of the 'Damascus' blade, from Persia or India, with its strange, textured 'damask' surface (Fig. 7.21). How was it done? What is the significance of such patterns?

The first objective description of the techniques of the mining, refining and working of metals was another of the great printed books of the Renaissance. *De re metallica* (*The Metal Trade*) was the posthumous work of a German physician, Georg BAUER (1494–1556), who translated his name into Latin – AGRICOLA. This work, with its evocative woodcuts of men and machines at work, gives a marvellous picture of the technology of the day. But he was mainly interested in more precious metals and said little about iron beyond a brief description of the basic smelting process (Fig. 7.22).

Metals became of interest in the seventeenth and eighteenth centuries for their chemical properties, but very little was said about their physical texture or structure. There were a few studies of solidification and fracture – for example, RÉAUMUR (1683–1757) noted the 'graininess' of

7.22   Smelting iron. From Georgius Agricola, *De re metallica*, Book IX.

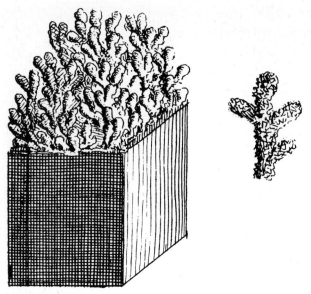

7.23　Fracture of grey cast iron, as seen under the microscope.

the fracture of cast iron (Fig. 7.23). In the early nineteenth century there were attempts at copying the various steel textures obtained by traditional craftsmen, and speculations about the 'grains' of which the material was supposed to consist. Careful chemical analysis also detected minor impurities, which could be correlated with the physical properties. By the 1850s the engineers who were fabricating iron and steel on a large scale for bridges and railways had a very good empirical knowledge of the strength of the materials that were available and there had been many new inventions, such as the Bessemer process for the smelting of ores and the production of steel from crude iron. Yet there was still no basic understanding at all of their familiar mechanical properties.

For the sources of this understanding, we need to follow the progress of other sciences, such as mineralogy. The observation of large natural crystals (Fig. 7.24), and the discovery of the geometrical laws governing their shape by HAÜY (1743–1822), naturally suggested theories of atomic lattices, but these were not immediately applied to metals. 'Crystals' of iron were sometimes observed inside hollow castings by iron workers (Fig. 7.25), but these seem to have been thought of as special forms, not characteristic of the material in its ordinary state.

Again, the *etching* of metals by acid solutions was a perfectly familiar phenomenon. It had been used for the decoration of armour from the

7.24   A form of calcite.

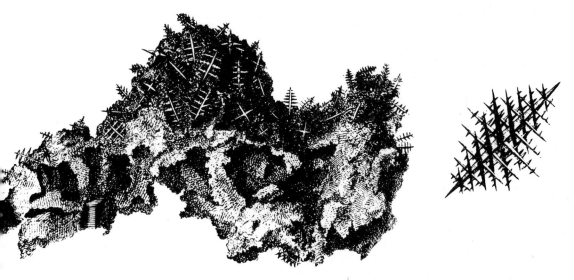

7.25   Group of crystals found in a shrinkage cavity in an iron casting.

sixteenth century (Fig. 7.26), and was, of course, the process used in making artistic plates for printing. The Japanese and Damascus swords owed their visible texture to carefully controlled etching. The patterns reproduce the slightly different solubilities of the various layers of metal that had been folded and wrought together, or had been differently heat treated, or had different chemical composition. But the connection of this delicate phenomenon with the underlying structure of the material was not imagined.

These ideas were brought together by Henry Clifton SORBY (1826–1908). For the benefit of Marxists, let me admit that he lived in Sheffield, and although he was not himself a steel master he was an amateur scientist whose wealth was derived from a family steel firm. In 1864 he did something quite simple, which could have been done at any time during the previous two centuries: it is a wonder, indeed, that it had not been one of HOOKE's famous discoveries (p. 42). SORBY took a piece of steel, polished it very finely, etched it with a mild acid – and looked at it under a low-powered microscope. He could see a complex pattern of areas of different shades of colour, which he could immediately identify

7.26   Detail of the breastplate of the armour of the Duke of Brunswick, *c.* 1540.

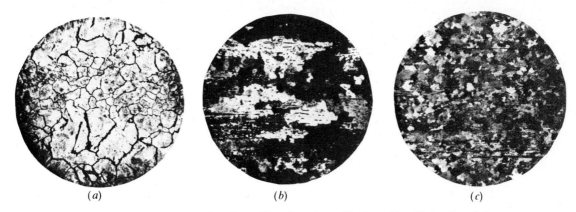

7.27 Sorby: (a) photomicrograph of Bowling bar iron (Sorby 1864. Reproduced from Woodbury type accompanying paper of 1887), (b) photomicrograph of wrought-iron armour plate, (c) photomicrograph of blister steel. Longitudinal sections, ×9.

as sections of small crystals of varying shapes, in various orientations (Fig. 7.27). He could also observe more localized features which could easily be explained as regions of chemical impurity or as precipitates of chemical compounds such as carbides of iron.

This was the starting point of the technique of *metallography* which is basic to the science of metallurgy. Having observed the basic crystalline forms one can build up theories in which the bulk metal is treated as an agglomeration of small crystals, within each of which the atoms are arranged on a regular lattice. The mechanical properties of the material, such as fracture, creep, plastic flow etc. are then connected with the forces between the crystallites, and their relative motion under stress. By studying the effects of heat treatment, cold working, alloying etc. on the crystalline texture, the metallurgist can begin to explain the corresponding effects on the mechanical properties. In other words, a rational science of the strength of metals is made possible, going far beyond mere empirical rules and tables of data.

Notice, however, that SORBY is not mentioned in Asimov's *Biographical Encyclopaedia of Science & Technology*, although he also made fundamental technical contributions to the microscopic study of minerals. This sort of research, lying outside the conventional boundaries of each of the classical disciplines of physics, chemistry and geology (although occupying precisely the interstitial region between them!) is evidently too 'applied' to be recognized as scientifically important. In fact the whole question of the microscopic structure of materials was given a cold shoulder by academic science until the last few decades.

Newman & Searle's *Properties of Matter* (1928) written for 'physicists', of course, scarcely mentions crystals; it would seem that they would have the student believe that 'matter' is always perfectly homogeneous and isotropic right down to the atomic scale. Metallurgy itself grew up along craft lines. The student was given a smattering of physics and chemistry, and then had to acquire a vast amount of apparently arbitrary knowledge about the metallographic texture of typical alloys, phase diagrams, tests of hardness etc., to fit him for the eminently practical life of a technical expert in a steel works or zinc smelter.

Since the Second World War metallurgy has gone through a second technical revolution as a result of the study of the typical defects in crystals – 'dislocations', 'vacancies', 'stacking fault' etc. – using X-ray diffraction and electron microscopy (Fig. 7.28). As is so often the case, these revolutionary concepts and techniques came from other academic 'disciplines' or were produced by people whose training was not originally in metallurgy. As a result, we now have reasonable semi-quantitative explanations of how and why certain types of steel will bend and break. Yet there are still very few cases where one can predict this behaviour in advance of the observation or, even better, where we can 'engineer' a new material for a specific purpose. In the meantime, the past century has seen enormous developments in the discovery of new alloys, with an extraordinary range of properties, for aircraft structures, turbine blades (p. 185), machine tools (Fig. 7.29) etc. This technical progress has been achieved very largely by conventional trial and error, based upon general metallurgical principles and specific knowledge, but inspired by shrewd intuition rather than deductive analysis. It is very hard work to build up a fundamental science that can keep pace with the actual developments of technique under the pressure of industrial or military demands.

Metallurgy itself, as a separate technical specialism, is being sub-merged within the new discipline of 'materials science', which includes the study of other materials such as ceramics and polymers. In the modern academic jargon it is referred to as an 'interdisciplinary' subject or as a branch of applied physics and/or chemistry. At the sublime level we refer to intellectual progress which permits the integration of various observations and theories into a coherent pattern giving an 'overview of the current state of the art', and we encourage cooperative research by an

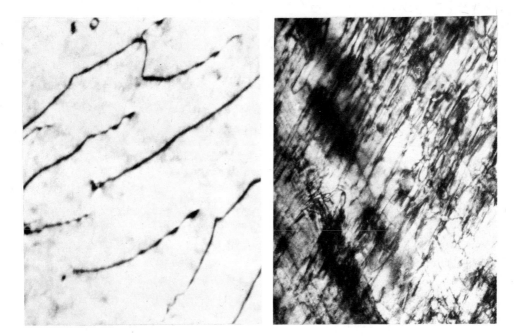

7.28 Low density dislocation array in fcc austenitic iron, as seen by transmission electron microscopy. ×70,000.

7.29 Very high density dislocation array in bcc martensitic iron. The high dislocation density gives the martensitic iron alloy its high strength. ×70,000.

interdisciplinary team of specialists drawn from the disciplines of solid state physics, physical chemistry, crystallography etc. In private it must be admitted that this change of name, and the overt affiliation with more prestigious academic departments, has helped to win new friends and new money for a subject that often seemed a bit too grubby and unglamorous for the adventurous student. We observe, in fact, a typical social phenomenon, whereby an applied science tries to borrow a cloak of respectability by becoming more theoretical, pure, and apparently useless. Many branches of medicine, such as the sub-specialities of anatomy, physiology and pathology, have followed the same path. Yet the skill of the experienced practitioner is still essential in the real world – the pinch of this or that ingredient, the tap of the hammer here or there, the design that looks right, the intuitive diagnosis – and these cannot be immediately explained or predicted by existing fundamental theory. The responsibility for teaching professional expertise still lies at the heart of every useful art, however far it may have advanced towards

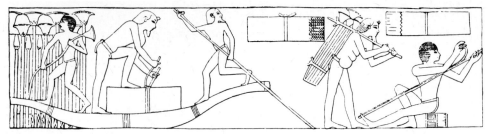

7.30 Collecting and preparing the papyrus plant. From a tomb at Thebes, *c.* 1500 BC.

the rainbow end of 'scientific medicine', 'scientific engineering', or 'the science of materials'. This applies just as much to the manipulation of dead matter as to an obviously more uncertain craft such as medicine or agriculture.

PAPER MAKING

As a final example let us consider the ancient craft of *paper making*. The name for paper comes from the Egyptian word for the reeds of the Nile – *Papyrus*. This material, which was used throughout the Graeco-Roman world, was made by cutting the stalk of the reed into thin slices, which were laid flat, crosswise like a woven mat, then soaked in water, pressed and dried in the sun (Fig. 7.30). It was in many ways quite a satisfactory material for manuscript writing, but was produced almost exclusively in Egypt. In the third or fourth century it was superseded by *vellum*, a fine parchment made from the skin of a calf, kid or lamb. This was washed, limed, unhaired, scraped etc., then stretched on a frame, where it was again washed, scraped, shaved, dusted, rubbed and dried (Fig. 7.31). Since the raw materials were available anywhere in Europe and since the processes were all hand labour, it could be manufactured locally – an important consideration in the subsistence economy of the Dark Ages and the Early Middle Ages.

The invention of true paper took place in China, somewhere between 200 BC and AD 100. The basic process is to cut up any natural materials containing cellulose fibres – for example, old cotton rags – and beat it into a slurry with water. This is spread in a thin layer on a mesh, where the water drains off, leaving a felted mass of tangled fibres that dries into a solid mat. To improve the texture and colour, or to get a good surface, the material may be treated with various sizes, filled with fine clay, bleached, dyed, smoothed etc.; but modern paper is not essentially different from the original Chinese product.

Paper came to Europe, via the Arabs, in the twelfth century. A period of 1000 years for the diffusion of even such a useful invention is not unusual. Yet it is instructive to ask ourselves whether it could have been hurried. Suppose, for example, that one were a determined and inventive inhabitant of Byzantium in the fifth century, who had been told the above outline of the paper making process: how much effort would have been required – the perfection of technique by trial and error, problems of reproducibility of quality, problems of production and marketing – to set up a paper factory in competition with papyrus and vellum. The evidence is that the invention travelled slowly by the diffusion of skilled craftsmen carrying a traditional method in all its details, which could be adapted to local circumstances and established step by step along its path. Techniques, also, travel around inside people!

It is also instructive to consider the role played by paper in the invention of printing. Without a cheaper material than vellum for the making of books, would GUTENBERG's invention (1450) have seemed economically profitable? Or should we ask how long it would have taken, under the stimulus of demand for a cheap material to feed the printing press, to invent paper independently in Europe? These are the sorts of

7.31 (Centre) Stretching parchment on a frame for drying. The workman shaves it with a half-moon knife which is of a special type, the blade being at right-angles to the handle. From a twelfth-century German manuscript. (Right) Smoothing the finished sheet with pumice, (left) cutting the sheet with the help of ruler and set square. Both from a thirteenth-century German manuscript.

7.32 The first illustration (1568) of a paper-maker at work: (at rear) water-driven stamp mill; (centre) press for squeezing sheets; (foreground) vat-man using the mould, while his boy carries away finished sheets.

questions that are asked about modern technologies with scarcely more hope of satisfactory answer.

By renaissance times, at all events, paper making was a standard guild craft, done by hand with simple machines. The elements of the process are easily recognizable – the beating mill driven by a water wheel, the screen and the drying press (Fig. 7.32). The main development, until the end of the eighteenth century, was the invention (1670) of the 'Hollander' which is rather like a fixed lawn mower for cutting up rags (Fig. 7.33). There were other improvements, for example, in the design and construction of the wire mesh screens, but these were all practical, piece-meal changes of technique which had nothing to do with any literate science. In the nineteenth century new chemical processes were discovered for releasing cellulose fibres from straw, wood pulp etc., but here again the inventions were made by simple experimentation with various recipes, and did not apply sophisticated chemical ideas.

From about 1800 onwards the manufacturing process was mechanized, with large-scale machinery (Fig. 7.34). The screen became an endless

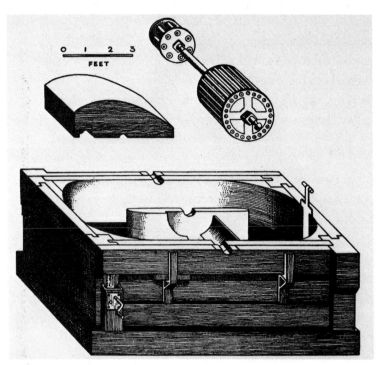

7.33 The Hollander.

7.34 Large paper making machine (modern). The Verti-Forma at Bowater's Mersey Mill, Ellesmere Port.

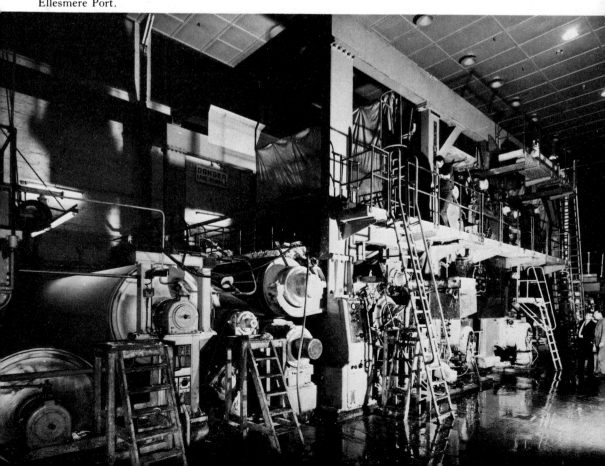

belt of wire mesh, on to which the pulp was fed in a continuous layer and drawn through a succession of pressing, drying, calendering rollers until it could be rolled up in ton loads at the other end. The invention, design, construction, and satisfactory control of these machines is a miracle of mechanical engineering, hydraulics, chemical engineering, electronics etc., to which many thousands of extremely ingenious people have contributed. Paper is a major product of modern society, being manu-factured, transported and used to the tune of tens of millions of tons each year. The expert skills required in this vast industry are taught in universities and technical colleges, or acquired by long experience in the factory or the field.

Is there, nevertheless, a 'science' of paper? What would be its ingredients? A practical knowledge of materials and methods, machines and processes would not be sufficient. To get down to fundamentals, we should have to start with the *biochemistry* and *plant physiology* of wood fibres. Then we would study the management of forests – *silviculture* –which would involve quite a lot of *botany*. The *chemistry* of extrac-tion processes, and the *mechanical engineering* of rapid automatic machinery are essential – but one must not neglect the *physics* of fibres, light absorption, colour, strength etc. The *economics* of distribution and marketing for such a large industry cannot be neglected – nor should we be ignorant of the technology of *printing*, which takes us into the *sociology* of the mass media – and so on.

This preliminary list illustrates the very general nature of what are usually thought of as highly specialized technologies. We cannot study the essential characteristics of such a material without impinging on many 'pure' sciences and also many important features of society at large. For example, paper making is closely connected with problems of environmental ecology, and with the recycling of waste products. Taught in this spirit, it could be a form of liberal education, as broad and 'relevant' as the direct study of sociology or the classics.

Even if we were to fasten our attention upon research on some very narrow technical problem, such as making paper stronger, we should need to bring together expert knowledge from several scientific discip-lines. Such a question requires information about biochemical struc-tures, about the physiological function of the fibres, about micro-meteorological conditions during the period of growth, and so on. An interdisciplinary approach is absolutely essential if we want to construct a scientific basis for an existing technology.

Note also that the craft origins of this technology would emphasize the scientific improvement or modification of the existing basic material, whereas one could envisage the invention of a completely new type of material to perform the same functions. We have seen this with polythene replacing paper as a wrapping material. Such an innovation is more likely to come by chance from some independent source, or perhaps from a generalized 'Thin Film Materials Laboratory', than directly out of a 'Paper Research Institute'. A highly developed technology has its own form of basic conservatism. Too much science directed too closely to the narrow end of maintaining a particular industry may even be a bar to technical progress. Buttressed by the vast inertia of large-scale social institutions such as factories and industrial corporations, a 'craft paradigm' can be as heavy to move as any 'theory paradigm' about which the philosophers contend.

8 INVENTION, RESEARCH AND INDUSTRIAL INNOVATION

Whatever is attempted without previous certainty of success, may be considered as a project, and amongst narrow minds may, therefore, expose its author to censure and contempt; and if the liberty of laughing be once indulged, everyman will laugh at what he does not understand, every project will be considered as madness and every great or new design will be censured as a project. Men, unaccustomed to reason and researches, think every enterprise impracticable, which is extended beyond common efforts, or comprises many intermediate operations. Many that presume to laugh at projectors, would consider a flight through the air in a winged chariot, and the movement of a mighty engine by the steam of water, as equally the dreams of mechanic lunacy; and would hear, with equal negligence, of the union of the Thames and Severn by a canal, and the scheme of Albuquerque the Viceroy of the Indies, who in the rage of hostility had contrived to make Egypt a barren desert, by turning the Nile into the Red Sea.

Samuel Johnson

The claim is often made that modern industry depends on basic science for its supply of innovations. More soberly, it is asserted that the support of pure science is justifiable because it will lead, eventually, to economic benefits through improved industrial products or processes. How far is this true?

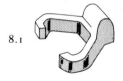

8.1

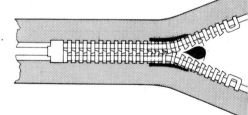

8.2   Zip fastener.

THE ZIPPER

This little piece of metal (Fig. 8.1), which is usually between 2 and 5 mm across, is made carefully and accurately in enormous numbers. It is, of course, one of the teeth of a 'zip' fastener (Fig. 8.2) – a great boon to everyday life. The *idea* of such a device is due to W. L. JUDSON, an

American mechanical engineer who applied for the first patent in 1891. It was a unique invention. No similar idea has been found in any of the patent files before this date. Over the next twenty years, the company formed to exploit the invention tried to make and market a saleable product. It was gradually improved, but they must have lost a great deal of money on it, for it was not really practical in its basic form. However, the 'Automatic Hook and Eye Company' took on a Swedish electric engineer, G. SUNDBACK, who in 1913 came up with the essential design of the modern fastener, and who also constructed machines to produce the parts and to attach them to the tape. But the clothing manufacturers were not interested until 1918, when the company got a contract to put their fasteners into flying suits. It was not until 1923, when fasteners began to be used on galoshes, that the new device was really accepted by the public. This little story has nothing to do with 'science', but has much to say about other factors in industrial innovation, such as individual inspiration, patient development and improvement of an imperfect product, the availability of finance for a risky venture, and the slow process of 'market penetration' against established techniques.

THE JET ENGINE

This is another class of mechanical invention. In January 1930 a British patent (Fig. 8.3) describing a proposed scheme for a gas turbine for the

8.3   Drawing illustrating British Patent No. 347,206: 16 January 1923.

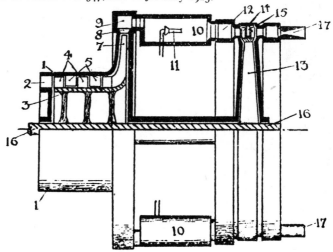

8.4 First model of the first experimental jet engine.

8.5 Combustion test arrangements outside the British Thomson Houston Factory, as used before the completion of the experimental engine.

jet propulsion of an aircraft was filed by Frank WHITTLE (1907–) then aged 23, who was in training as an engineer in the Royal Air Force. His basic idea was to generate the propulsive jet of air within the gas turbine itself, rather than using the engine to drive a propeller. But not much came of this idea until 1935, when WHITTLE was at Cambridge studying engineering. He talked to a few people about it, and a small company was set up (with £2000 capital) to exploit the invention.

The first model (Fig. 8.4) was built in 1936 on the premises of a big engineering firm, the British Thomson Houston Company. From this time on, WHITTLE could not be called a 'lone inventor': he had a small team of assistants, some modest engineering facilities, and some interest from the RAF and other aeronautical organizations. Nevertheless the whole effort was obviously very makeshift and done on the cheap (Fig. 8.5). Up to 1939 the total expenditure involving the design and construction of a series of models was no more than £20,000 – a sum that would scarcely pay for the cups of tea drunk during the design of a modern jet engine. There were many failures (Fig. 8.6) and technical difficulties with the design of turbine blades, combustion chambers etc. On the outbreak of the War, a more concentrated effort began with much larger resources, but it was not until 1942, six years after the first attempt,

8.6   Failure of turbine disc of third model of experimental engine (February 1941).

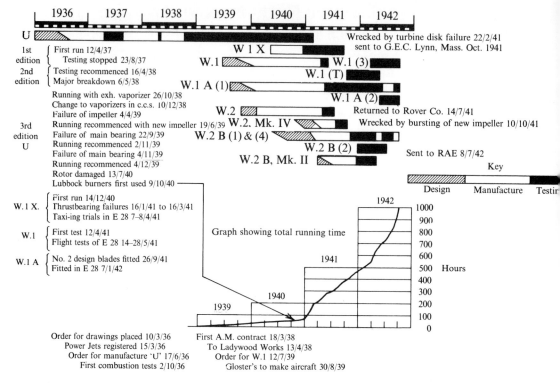

8.7 Whittle jet. Diagram illustrating the principal events in the early history of the jet propulsion gas turbine.

that an engine was built that was reliable enough to be put into production (Fig. 8.7).

This is a well-known story, told with vigour by WHITTLE himself. It illustrates many significant phenomena.

We notice, immediately, the importance of the man with an idea, and the will to put it into practice. WHITTLE received a good deal of expert assistance at various stages but he was the leader of the development team until it was successful. This innovation was not the product of an 'industrial research laboratory' in a large firm. Indeed the established British aero-engine manufacturers were not interested in this crackpot idea. The concept of a gas turbine had been proposed already several times, but they 'knew' from experience that it would not work. WHITTLE had to fight hard to break existing prejudices which were incorporated in the 'engineering science' of his day.

On the other hand, WHITTLE was a thoroughly well-trained professional engineer, able to call on the best advice and experience. He used the theoretical ideas of thermodynamics (see p. 26) and aerodynamics to prove that his engine would work and to design it efficiently. In particular, by purely scientific arguments he demonstrated the necessity of shaping the turbine blades according to correct aerodynamic principles. This invention would have been impossible without the corresponding fundamental knowledge of physics and applied mathematics. And the engine itself could not have been made to work without the new alloys, capable of standing very high temperatures, that were just being developed by the metallurgists (see p. 172).

The influence of the War (chapter 13), giving urgency, money, men, and production facilities to the little team, is also typical. Whether or not we regard the jet aircraft as a blessing or a bane, it is doubtful whether it would have been developed with anything like the same impetus without this incentive.

Most significantly, it was not a unique invention. The British team were working secretly, knowing nothing of the parallel German development. Two students at Göttingen, Hans von OHAIN and Max HAHN, unaware of WHITTLE's patent themselves patented a similar system in the mid thirties, and several major German aircraft firms began work on it in secret competition. A Heinkel with a jet engine (Fig. 8.8) made its

8.8   First jet aeroplane; the Heinkel He-178.

first flight on 27 August 1939, nearly two years before the first British jet flight. The failure of the Luftwaffe to win air superiority with a fast jet fighter during the War was mainly due to incompetent organization at the top. The German Air Ministry believed in piston engines and failed to give adequate support to the brilliant engineering teams of their aircraft companies. It is clear, in this case, that the jet concept was actually implicit in the engineering knowledge and practice of the 1930s, and the development would have occurred, in any case, at about that time.

Nowadays, of course, the manufacture of jet engines is a huge international industry with a turnover of hundreds of millions of pounds a year. Along with this goes an enormous technology – i.e. technical knowhow gained by experience or accurate calculation. This skill is used continuously to improve the product in power and performance (Fig. 8.9). This is a very costly activity, but with a good, almost predictable economic return. When a firm such as Rolls Royce or Pratt & Whitney design a new engine such as the RB 211 (Fig. 8.10) they can calculate in advance the power and weight they expect to achieve, and have reasonable confidence that various unforeseen defects will be put right by research, minor invention or redesign as the development proceeds. It is scarcely necessary to remark that there may be a large factor of *commercial* risk in such a process, but the technical goals are

8.9   Improvements in the performance of by-pass jet engines. Fuel consumption (vertical scale) and weight have been halved by technical developments over a period of twenty years.

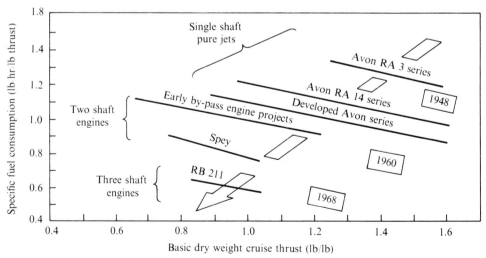

8.10 RB 211 engine.

usually reached with uncanny accuracy. Are such transformations to be regarded as genuine 'innovations'? Perhaps with a closer look at each part of the engine, we should observe, at a 'microscopic' level, much the same inventive skill and patient improvement as in the invention and development of the machine as a whole. Those responsible would be using the same sort of mixture of pure science and craft experience as WHITTLE himself: it is a matter of taste whether one calls them 'scientists', 'engineers' or 'technologists'. To complete the cycle, we should almost certainly discover within the industry an established technological orthodoxy, such as the practical skill of the big piston engine manufacturers in *their* day of supremacy, which would stand in the way of fundamental innovation.

Once established, a new technology may be self-sustaining in the supply of innovations. For example, the number of railroad patents registered each year in the United States parallels very closely the annual output of rails (Fig. 8.11). In a capitalist economy investment in an industry apparently generates the technical inventions and innovations to make better use of the money raised. The causal connection suggested by this fascinating graph is obviously not proven, but it does indicate that technology is to a considerable extent autonomous for its sources of change, and is largely independent of pure science, which merely provides a climate of basic ideas and general principles.

PENICILLIN

To redress the balance in this debate, let us now consider a technical innovation that undoubtedly sprang directly from pure scientific research. The accidental observation of the effects of the mould *penicillium* on a bacterial culture is famous (Fig. 8.12). It was fortunately observed by the right man. In 1928 Alexander FLEMING (1881–1955) was working in a small but high quality laboratory of bacteriology at St

8.11 Railroad rails: output and patents, USA 1860–1950.

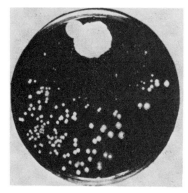

8.12 Original slide of mould growth of *Penicillium notatum*.

8.13 Fleming in his laboratory.

Mary's Hospital, London (Fig. 8.13). He was deeply interested in the bactericidal or bacteriostatic action of natural products, and immediately began detailed research on the *penicillium* phenomenon. Over a period of three or four years he did an excellent job of testing the action of the extract from the mould, showing that it was non-toxic to animals and humans, concentrating it etc. The results of the research were duly published: but FLEMING was a taciturn, inarticulate man with no flair for publicity, and the discovery was almost completely ignored. He was not himself able to command the biochemical and chemical resources that were needed to isolate and purify the active agent, and in the commercial pharmacological world it was generally considered that the concept of a non-toxic bactericide was a contradiction in terms.

8.14 Part of the plant for penicillin extraction in which 10-gallon milk churns were used as mixing vessels and garden fountain pumps for transferring the liquids from one part of the plant to another.

8.15 A battery of 10,000-gallon fermentation tanks at Glaxo Laboratories plant at Ulverston.

Nothing was done, therefore, for another ten years. Then in 1939 a group under Howard FLOREY (1898–1968) and Ernst CHAIN (1906–) at the Pathology Department at Oxford began work on natural antibiotics. They came across FLEMING's paper in the literature (they took no steps to consult him, thinking he was dead!) and soon found penicillin to be highly effective and capable of concentration and purification (Fig. 8.14). By 1940 they had sufficient material to demonstrate its therapeutic power on a policeman with acute blood poisoning – but they had not enough penicillin to complete the treatment and the patient died.

There were still enormous difficulties in actually preparing the drug on a useful scale (Fig. 8.15). FLOREY went to the United States in 1941 and with the needs of warfare as a powerful lever was able to harness the large resources of the American pharmaceutical industry to the task of development and production. It took six months to get enough of the active penicillin to treat one case: in eighteen months they had enough for 200 cases: after thirty months processes were available which could be used for large-scale industrial production (Fig. 8.16). For all this,

8.16   Graph of US penicillin production 1943–50 (assuming 10^7 units equivalent to 1 medical treatment.)

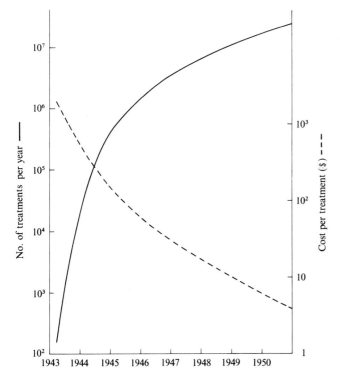

8.17 Fleming receiving the Nobel Prize.

FLEMING, FLOREY and CHAIN were awarded the Nobel Prize for physiology and medicine in 1945 (Fig. 8.17), and received knighthoods etc. Being pure scientists (although in fact men who cared deeply about the applications of their discovery) they were, no doubt, well satisfied with the honour and glory. The industrialists had to solace themselves with the financial profits of the operation.

Once again, from this small seed has grown an entirely new industry on a very large scale. Penicillin itself is now only one of a whole range of antibiotics derived from micro-organisms. The technology of the manufacture, testing, distribution and sale of these products is yet another highly specialized and differentiated human skill, which has itself been improved and transformed by many minor innovators.

This field has gone through an interesting and important stage of secondary discovery. Because of the great difficulty and expense of the biological production of penicillin, an enormous scientific effort was put into establishing its chemical structure, in the hope of discovering a path for non-biological synthesis. The summary of this research runs to 1000 pages. This structure (Fig. 8.18) was elucidated in 1949 by DOROTHY HODGKIN (1910–), and by an American team. Should we count this

8.18 Penicillin – formula.

an enterprise of 'pure' science or was it just 'applied' research? It certainly used to the full all the skill, experience and intuition of the academic organic chemist, and was confirmed by the intellectually refined techniques of X-ray crystallography.

Complete synthesis of penicillin was found to be practically impossible, but CHAIN wanted to go on. He argued that it was essential to set up a team of chemists, biochemists and bacteriologists, with a small-scale production plant to study the process in detail and explore its potentialities. He did not get the support he needed in Britain, from Oxford University or from the Medical Research Council, so he accepted the post of director of a chemical microbiology institute at Rome, where he was given much better facilities. However, the Beecham pharmaceutical group then decided to go into research on penicillin, and built an excellent laboratory to which CHAIN became a consultant. The result soon was the discovery of a new group of penicillin derivatives with substituted side chains (Fig. 8.19) which could be taken orally and which

8.19 Ampicillin – formula.

were effective against bacteria that had grown resistant to ordinary penicillin. This work cost Beecham's about £1m annually but the new products were eventually being sold at a turnover rate of £10–20m. It is an excellent example of *directed* research, coupled with willingness to take quite a big financial risk on the outcome. This sort of account does not, of course, include the very large balancing entries for gambles that did *not* pay off, despite considerable investment of time and money in research and development.

For another product of the industrial research laboratory let us consider *nylon*. Many peculiar 'plastics' with practical uses had been made by accident in the nineteenth and twentieth century – vulcanized rubbers, celluloid etc. Until about 1925 it was not at all clear how these should be characterized chemically. They were thought to consist of small molecules linked up in irregular and untidy ways. In 1926 Hermann STAUDINGER (1881–), Professor of Chemistry at Freiburg, began to study these materials systematically, and showed that they mostly consisted of large molecules in which the atoms or chemical groups were arranged in very long chains. For this significant discovery in pure chemistry, which is the beginning of the modern science of *polymers*, STAUDINGER won the Nobel Prize in 1953. It was the key concept in the understanding of both natural and artificial polymeric materials and in the deliberate production of new materials with desirable properties.

In 1927 E. I. du Pont de Nemours & Co., a large American corporation mainly specializing in the manufacture of explosives, decided to start a programme of fundamental research. They budgeted $250,000 a year and hired an excellent Harvard chemist, Wallace H. CAROTHERS (1895–1937) then aged 32, to direct the research. CAROTHERS became interested in the fundamental characteristics of polymerization. In particular, he observed a new phenomenon: a fibre could be drawn from a hot molten polymer, and strengthened and made more elastic by further cold drawing. He experimented with various materials, and was pressed to find a commercially useful compound. Turning his attention to the *polyamides*, which are relatively simple chemically (Fig. 8.20), he

$$NH_2.R.NH_2 + HOOC.R'.COOH \rightarrow NH_2.R.NH.CO.R'CO.NH.R.NH.CO.R'...R.NH_2$$
diamine   dibasic acid   polyamide

8.20 Nylon polymerization reaction.

found that these had excellent potentialities as fibres. Armed with STAUDINGER's basic concepts, and with his own fundamental grasp of the effects of molecular shape and reactivity of the constituents on the polymerization process, he was able to guide the research towards the

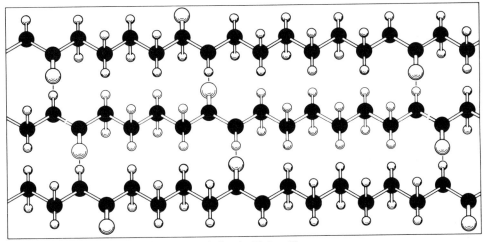

8.21   Cross-linking between polymer chains in Nylon 66.

best possible combination in this group of materials. For example, the melting point of the polymer depends to some extent on the pattern of cross-linkages between neighbouring chains (Fig. 8.21). This, in its turn, depends upon the relative numbers of carbon atoms in the two different constituents of the polymer (Fig. 8.22). Thus *Nylon 66*, the final optimum fibre material, has six carbons in each of the two sections.

8.22   Variation of melting point as a function of the index of hydrogen interactions for polyamides with even numbers of CH₂ groups.

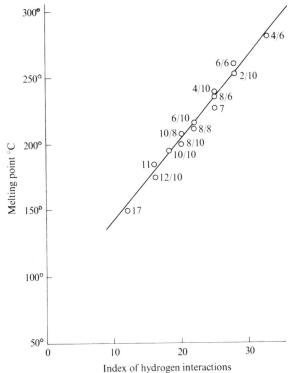

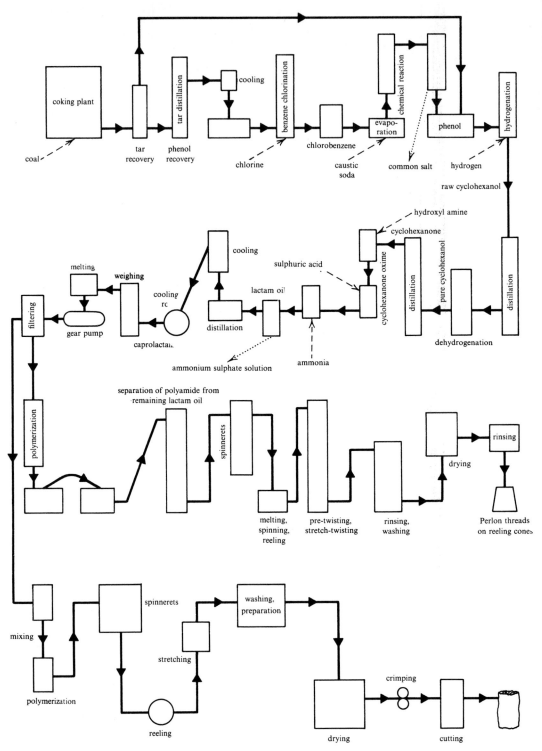

8.23 Nylon plant, schematic drawing.

Nevertheless despite relatively early success in finding a very satisfactory material, large-scale production of nylon fibre did not begin until 1939. It had cost $6m for research and $21m for plant development over a period of eleven years. It was necessary to work out the chemical processes for the production of the basic components in bulk – a typical expert's task in the highly skilled art of chemical engineering. It was also necessary to develop a whole new range of machinery to draw, stretch and spin this unusual fibre (Fig. 8.23). Nylon, like most significant industrial innovations, can be said to be the product of the labour of many hundreds or thousands of people, working in a large team under systematic management. It is a classic example of big-firm research and development with a long-term commercial aim. This is what people have in mind when they say that innovation nowadays is mainly the product of the industrial research laboratories of large corporations. Remember, however, two essential factors – the 'academic' science of STAUDINGER, which provided the conceptual structure within which the experimental observations could be comprehended, and the brilliant leadership of CAROTHERS who knew how to manipulate the new concepts to 'engineer' a material with desirable properties.

At this stage we ought to consider at some length the relationship between a research and development organization and its parent body. Within an industrial corporation this can often be very subtle and sometimes unhappy (Fig. 8.24). There are pressures from the research side to support everybody's pet notions, or to keep alive projects that are not going well. From the production and sales divisions come demands for new products with at least an immediate market, or for short-term trouble-shooting on the factory floor. The research workers are said only to care for getting publications in the academic journals whilst the management only look at the current balance sheet. Meanwhile the path of a bright idea from the research department to the production plant is strewn with boulders. It is an area for many a gladiatorial combat and provides excellent bulky material for Ph.D. theses in sociology. In fact it is too big a subject in its own right – and would eventually be found to illustrate the whole theme of this book in a microcosm.

8.24 Cartoon from *International Science & Technology*, New York

But we must surely say something about the largest, most famous and most successful of such institutions – the Bell Telephone Laboratories of New Jersey founded in 1925. This vast organization is supported out of the income of the American Telephone and Telegraph Company, for whose benefit it is supposed to work. In fact it also provides a home for some of the finest pure scientists in the world, who contribute as much as any great university faculty to the academic literature in physics, applied mathematics, electronics etc. From time to time other mammoth industrial corporations have tried to copy 'Bell Labs' by hiring large numbers of pure scientists and telling them to spin away at their own little webs – but the results have been disappointing, in terms of profits, over five or ten years; eventually the accountants have had their say, and everybody is told to get back to applied research on the company's traditional products.

By some means, Bell Labs seem to get first-rate applicable and developable results out of their research teams. The invention of the

8.25   Central telephone office battery plant with 1680 AH plastic jar cells.

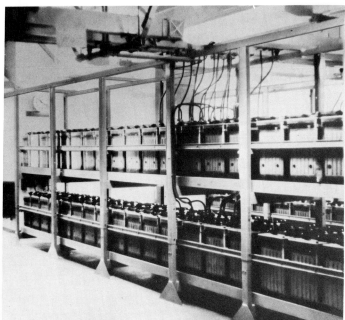

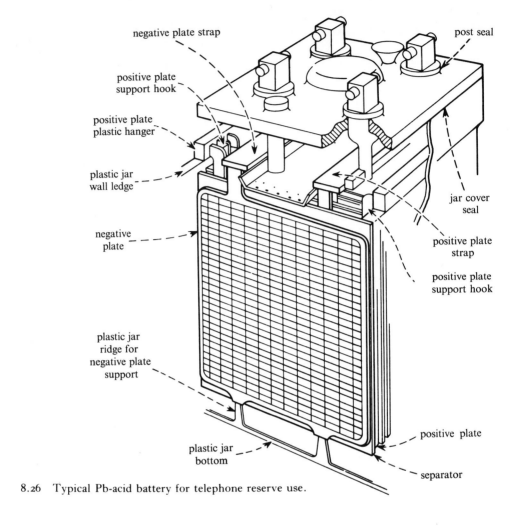

negative plate strap

post seal

positive plate
support hook

positive plate
plastic hanger

plastic jar
wall ledge

jar cover
seal

negative
plate

positive plate
strap

positive plate
support hook

plastic jar
ridge for
negative plate
support

positive plate

plastic jar
bottom

separator

8.26   Typical Pb-acid battery for telephone reserve use.

transistor by BARDEEN, BRATTAIN and SHOCKLEY in 1948, starting from an objective study of basic semiconductor phenomena, is the most famous example in our time of a fundamental technical innovation arising directly from pure science. But a more typical example of industrial research and development is reported in the *Bell System Technical Journal* **49** (1970).

A large telephone system must have a 'float' battery of storage cells to maintain electric power in case of a failure of the mains supply. These cells are much like ordinary car batteries (Fig. 8.26) although much larger. They also have a different duty cycle, since they are seldom heavily discharged, but must be reliable, in emergency, over long periods. Unfortunately, experience had shown that the conventional batteries did not last for nearly as long as the telephone engineers would have liked, and tended to break down after ten to fifteen years even of this very sedentary life. They also started fires, due to leakage of acid, followed by short circuits, which could ignite the polystyrene jars.

The problem, then, was to design a completely new type of battery with a life greater than, say, thirty years. Although apparently marginal to these telephone systems, this is not a negligible economic problem. The company were replacing 100,000 cells each year. If these were costing $100 each (which seems a conservative estimate, considering their size) then the gain that might be achieved would be of the order of $5m per year. A substantial research and development project was fully justified.

The basic difficulty was well known and easily stated. In order to make a plate that is strong enough to support its own weight when hung from the top of the jar into the acid electrolyte, the lead must be alloyed with 0.05% of calcium. This impurity leads to corrosion and electrochemical distortion of the plates (Fig. 8.27) during those many years of idleness at

8.27 Corroded positive plate with original grid overlaid to indicate initial dimensions.

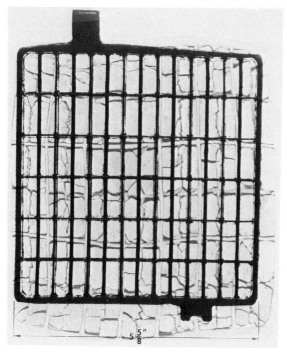

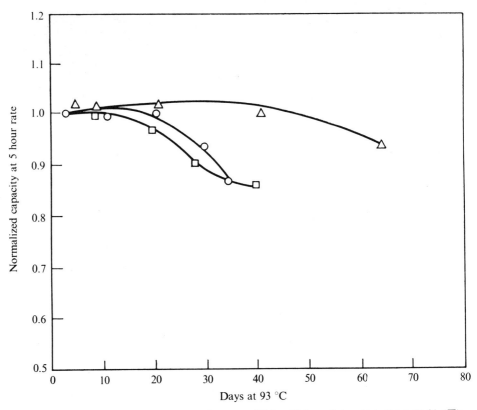

8.28　Normalized 5-hour rate capacities vs days at 93 °C for 5⅝ in × 5⅞ in × ¼ in thick PbSb (□), PbCa (○) and pure Pb (△) grids.

full charge. The first step was to check the supposition that pure lead plates would last much longer under the electrochemical conditions. This is not so easy to verify: one cannot wait ten to fifteen years to complete the experiment. It was a research in itself to verify that running the cell at 93 °C (i.e. near boiling point) hastens all the ageing effects uniformly, so that a run of a few days could stimulate the behaviour of a real cell over many years (Fig. 8.28).

But a pure lead plate is much too soft to be hung in the conventional manner. A completely new arrangement of the electrodes is needed to maintain the mechanical stability of the system. The plates must be stacked horizontally, supporting each other with spacers (Fig. 8.29). But a flat disc tends also to distort electrochemically, so eventually the designers hit upon the idea of making the plates into conical discs, which could be stacked neatly on each other in a cylindrical jar (Fig. 8.30).

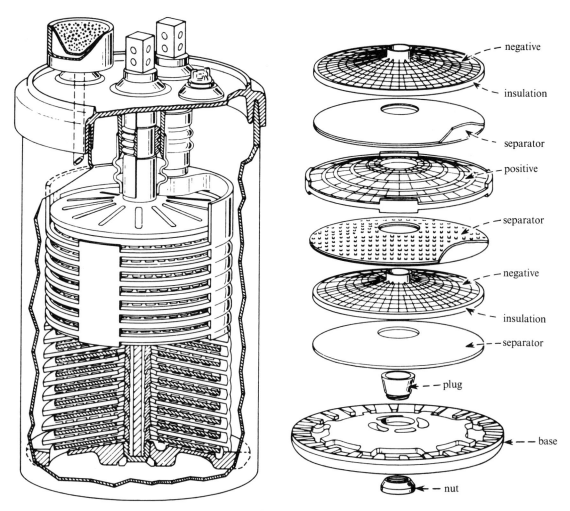

8.29   The Bell System battery – cutaway and exploded view.

This redesign of the cell structure is the most radical innovation made in this development; but many other scientific and technical problems had to be solved to obtain consistent long life performance. These include the following.

1. Shaping the *positive* grids to avoid distortion.

2. Inventing a new '*paste*' of lead oxide. It was found that a mixture of the constitution $4PbO.PbSO_4$ had much better mechanical properties than the conventional paste when converted electrically into PbO_2.

3. Study of the electrochemical behaviour of cells under 'float' conditions. This demonstrated the importance of *cell uniformity* in each battery.

8.30 Trial installation at Murray Hill, New Jersey, Bell Telephone Co. central office.

8.31 Injection-moulding machine.

4. Satisfactory *venting* of gases when the cell is overcharged.

5. The proper choice of jar and cover materials. These were eventually made of PVC (polyvinyl chloride), which is cheap, rigid, flame resistant, tough, and has high impact strength. But these jars weigh 17 lb each and are the largest single items ever moulded in PVC (Fig. 8.31).

6 The *chemical purity* of the plastic parts turned out to be very important. By keeping the electrochemically active impurities in the PVC etc. to a few parts in a million, it is hoped to reach a fifty-year life for the cell.

7 The *jar cover seals* were a weak point of the old cells. A new way of making these seals was invented (Fig. 8.32). The transparent lid of the jar is put on with an infra-red-absorbent layer round the lip. When this is irradiated from the outside with infra-red light, it heats up, melts, and seals the cover on to the jar. The joint is very strong and leak proof: this invention is likely to have applications in quite different industries, such as food presentation.

8. The *post seals* through which the terminals pass in and out of the jar

8.32 Jar-cover seal assembly and sealing sequence (schematic).

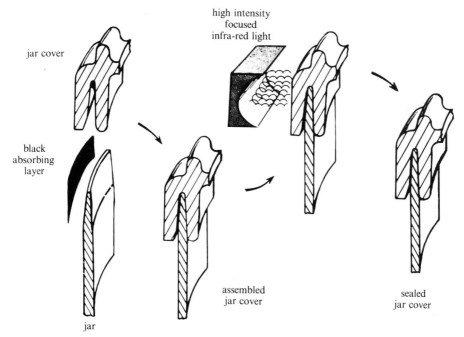

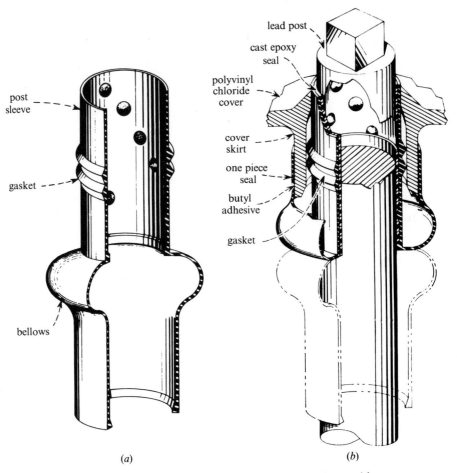

Labels in figure (a):
post sleeve
gasket
bellows

Labels in figure (b):
lead post
cast epoxy seal
polyvinyl chloride cover
cover skirt
one piece seal
butyl adhesive
gasket

(a)   (b)

8.33   Final post seal assembly: (a) one-piece post seal, (b) post seal assembly.

must allow movement of the lead post against the PVC cover, yet they must be impervious to strong sulphuric acid. The trick that was discovered was to put a thick butyl rubber sleeve over the post, to which it was sealed with epoxy resin, and then turn it back over a cylindrical tube passing through the cover (Fig. 8.33).

9. It was important to get good, *low-resistance joints* at the edges of the positive plates. New automatic machinery was designed to bond the plates together with molten lead.

10. A plastic *rack* system was designed (Fig. 8.34) into which the cells could easily be packed. This had to stand up to earthquakes, and, for some uses, to nuclear attack in a 'hard' site.

Let it be emphasized that the electrochemical process used in the new cell is exactly the same as in the conventional lead–acid battery. The research did not involve new basic scientific results. But the resultant of all these minor changes and technical innovations is a radically new product, which is, by any standards, more than twice as good as the old one. The new cells are now being produced by one of the independent battery manufacturers that supply AT & T, for the company itself is not permitted to manufacture its own equipment. Since they have done the research, and development, it is interesting to enquire how the patent rights will be assigned and royalties paid: this question is not dealt with in the technical articles from which this account is drawn. From the evidence of these articles, a team of at least twenty people were working on this problem, so it must have cost something like $½m to $1m per year over a period of five or six years. Evidently the research was successful, not only in creating a new and much improved device but in terms of profit and loss. This team, incidentally, contained about seven Ph.D.s; the remainder had B.Sc. or M.Sc. qualifications. They were mostly experts in electrochemistry and chemical engineering, but also included mechanical engineers, plastics experts etc.

8.34   New rack ceiling-suspended for service in a 50-psi hard site.

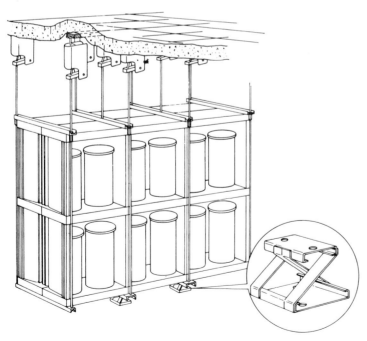

It is difficult, and dangerous, to generalize on these examples of industrial innovation and on the many other cases now collected by various authors. But a few comments are in order.

We observe, for example, that individuals of powerful mind or character often play an important part, whether as inventors or as entrepreneurs. But the 'idea' of an invention is only the beginning, and development to the stage of public sale takes a long time. The characteristic modern time lag of ten to fifteen years is not really so much shorter than in earlier centuries.

The real costs of industrial innovation arise in the development phase, where, in fact, the majority of the 'scientists' in industry are actually employed, usually in very large firms that can afford the immense resources needed to carry a scheme to completion. There are also many intellectual obstacles (i.e. prejudices) against new products, both from the manufacturer and from the customer. Efficient, imaginative marketing may be the key to success, rather than brilliant design.

These were success stories. But many similar ideas, that looked just as promising, failed at various stages. The commercial benefits of an invention have often gone to others than the prime discoverers. It is extremely difficult to construct a simple balance sheet of the profits and losses from any single invention or for innovation with any firm or industry. All that we know is that, when all failures are discounted, the overall *economic* return, for society at large, is enormous (see p. 263). The consequences of innovation, in other respects, are not for the moment in question.

In this activity, one of the dominant social processes of our time, 'pure', 'basic', 'fundamental', 'academic' science is seldom an immediate factor. But it is essential to innovation by providing the inventor, technologist or engineer with a training ground in the basic principles, and as the main source of new ways of thinking about old problems. To appreciate this role we must look back a long way, past the ten to fifteen year development phase, past the invention itself, to the previous fifty or one hundred years, when these principles will have been gestating.

Finally, it is important to emphasize the transfer of technical ideas *between* technologies – e.g. from plastics to electrochemistry, from high temperature alloys to mechanical engineering. The full story of many inventions often contains a significant episode when a man has moved from one technology to another, or from one company to another, and has been able to see things in a new light. Ideas move around inside *people*.

9 BIG SCIENCE

The hardest bones, containing the richest marrow, can be conquered only by a united crunching of all the teeth of all dogs.   *Kafka*

TELESCOPES

The instruments of scientific research have increased enormously in size, complexity and expense over the centuries. Astronomical instruments, of course, have always been large and expensive by the standards of their time. Tycho de BRAHE (1546–1601), as court astronomer to the King of

9.1   Sextants used by Tycho Brahe: (right) by Erasmus Habermel, Prague, 1600; (left) probably by Jost Bürgi, *c.* 1600.

9.2   The great mural quadrant from *Astronomiae instrumentae mechanica*.

9.3 Delhi Observatory, Jantar Mantar. Monumental astronomical instruments built by
Maharaja Sawai Jai Singh II, *c.* 1724.

Denmark, and then to the Emperor Rudolph II, spent very large sums
on his apparatus, which was beautifully constructed and often very
massive (Figs. 9.1, 9.2).

In India there still stand several great observatories built for the
Maharajah JAI SINGH, probably for astrological observations. Each of
the instruments (Fig. 9.3) is a massive structure of stone and brick, like a
great sundial 100 feet in height. But they could be used by a single person,
noting the movement of shadows or the transit of a planet on a very broad
scale. Astronomy and astrology were the household sciences of kings and
princes, and no expense need be spared to equip their practitioners.

The first telescopes were, as we have seen (p. 20), very long and
unwieldy. But in the late eighteenth century Sir William HERSCHEL
(1738–1822) began the construction of very large instruments which
revolutionized astronomy. His son, Sir John HERSCHEL (1792–1871)
went to Cape Town for five years in 1834 and set up an 18-inch reflector

9.4 Sir John Herschel's large reflecting telescope at the Cape of Good Hope.

9.5 Lord Rosse's 72-inch reflecting telescope at Birr Castle.

(Fig. 9.4) which he used to map the Southern skies, thus completing his father's work.

The grandfather of them all was the 6 foot reflector (Fig. 9.5) built at Birr Castle, Co. Offaly, Ireland, by the EARL OF ROSSE (1800–67). It cost him £30,000, which was an immense sum in 1845; but with the largest telescope in the world he was able to observe the spiral structure of the extragalactic nebulae. Incidentally this great aristocratic amateur was not without social conscience: when the potato famine ravaged Ireland from 1845–8, he abandoned his research and did all he could to combat distress and disease amongst the starving people.

But come forward another century, and look at a typical large modern telescope, such as the Newton telescope at the Royal Observatory (Fig. 9.6) built in 1967. The aperture, 100 inches, is not much larger than

9.6 Isaac Newton telescope at Royal Greenwich Observatory, Herstmonceux.

that of the Rosse telescope, but of course the whole instrument is far more complex. The structure is built to much higher precision, with beautifully controlled driving motors, and a range of auxiliary equipment that multiplies many times the power, accuracy, convenience, spectral range etc. of the observations that can be made. And this is only half the size of the famous Mt Palomar telescope.

For real size, we should turn to the radio telescopes. The Arecibo bowl (Fig. 9.7) is no less than 1000 feet across, built into a mountain, with a receiver slung on cables to move across the focal plane. Such an instrument is at the limits of engineering design. The demands of research are always for the maximum size, the closest tolerances, the highest sensitivity that current techniques permit. To build, maintain and use an engineering system of this size and complexity calls not only for money in millions of dollars, but also for organization and management on the industrial scale. This transformation of the scale of human operations in research is what we mean by the appearance of *Big Science*.

9.7 Arecibo Observatory, showing large radio telescope.

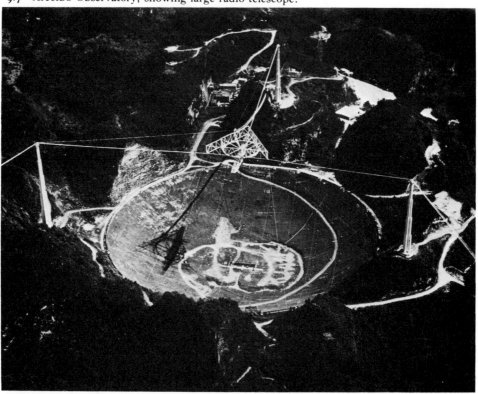

9.8   Advertisement, 1897.

9.9   Rutherford holding in his hands the apparatus with which the artificial disruption of the nucleus was first demonstrated.

PARTICLE ACCELERATORS

The biggest pure science grew out of atomic physics. Let us go back just one lifetime, to 1895, when Ernest RUTHERFORD (1871–1937), a bluff, energetic New Zealander, arrived at the Cavendish Laboratory, Cambridge (p. 62), to begin research under Sir J. J. THOMSON (1856–1940). The sort of equipment they used (Fig. 9.8) could be bought for a few pounds – not much money even in those days. Even when RUTHERFORD 'split the atom' in 1919 – i.e. produced a nuclear transformation in nitrogen by bombardment with alpha particles – the apparatus was made in the laboratory workshop by a skilled technician, and could be held in the hands (Fig. 9.9).

By 1930, however, the scale had increased in size and cost. The first

9.10 Cockcroft–Walton apparatus with which high voltage particles were first made to effect artificial disintegration of the atomic nucleus.

linear accelerator of nuclear particles to high energies (Fig. 9.10) was built by J. D. COCKCROFT (1897–1967) and E. T. S. WALTON (1903–). It cost the substantial sum of £1000 and occupied the best part of a room with its column of insulators to carry a voltage of several hundred thousand volts. Experiments with this device required the collaboration of several research workers, but were essentially simple and direct.

A year or so later E. O. LAWRENCE (1901–58) and M. S. LIVINGSTON (1905–) at the University of California began operating their first *cyclotron* (Fig. 9.11). This occupied about the same space as the COCKCROFT–WALTON accelerator, but gave particles of a million electron volts. They were lucky to find a big electromagnet, weighing 85 tons, that nobody wanted, so it only cost about $10,000 to build. This design was so successful that by 1939 it had been 'stretched' into a much

9.11   First cyclotron used in nuclear transformations.

9.12   Early photograph of 60-inch cyclotron.

9.13   General view of 184-inch cyclotron, Proton Con 152, which produces 340 MeV protons. The concrete shielding, partially removed in this photograph, is 15 feet thick.

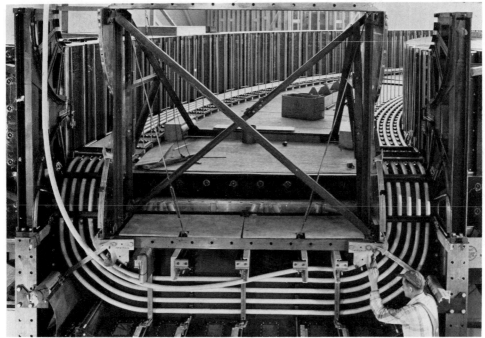

9.14   Coil winding of bevatron magnet (Bev 395).

bigger machine (Fig. 9.12), 60 inches in diameter, yielding particles of 12 million electron volts. But costs were now in the $100,000 range, and proper engineering design had taken over from the makeshift hand-made structures of the first apparatus. This machine required a special large room in a laboratory designed around it, but could still be operated by a few physicists and technicians working informally together.

The logic of high energy physics led inexorably to the next upward step in energy – the *184-inch synchrocyclotron*, planned before the war and built just afterwards, at a cost of $1.8m. This machine was now on the scale of, say, a big steam turbine for power generation, and could only be housed in a special building (Fig. 9.13) with facilities for auxiliary equipment, control panels etc. The proton beam had an energy of 340 million electron volts, and produced such copious quantities of dangerous radiation that the handling of the large blocks of concrete needed for shielding was itself a major engineering task.

The escalation of cost and size has continued. The successor to the synchrocyclotron at the University of California was the *bevatron*, built in 1953 at a cost of $9m. This enormous machine (Fig. 9.14) is 34 metres

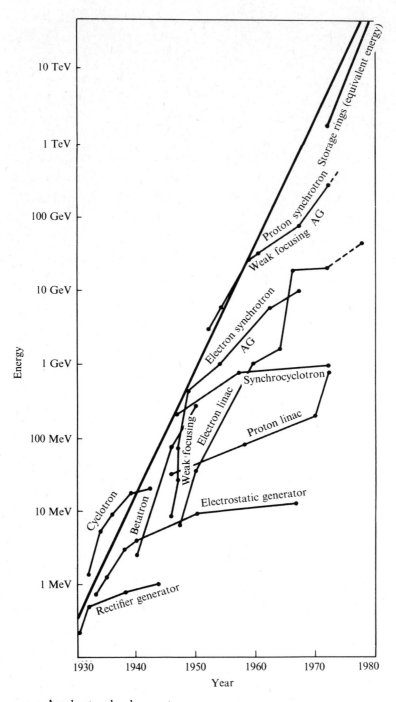

9.15 Accelerator development.

in diameter, and produces particles at an energy of 6.3 GeV – i.e. 6,300,000,000 volts. Together with the bubble chambers, beam handling magnets, and other apparatus needed to do experiments on elementary particles at these very high energies, this research instrument requires a whole complex of buildings on the scale of a factory or power station.

But the 'learning curve' of accelerator energy development did not stop in 1960 (Fig. 9.15). To go a step further, we should look at the CERN (Conseil Européen des Recherches Nucléaires) *proton synchrotron*, situated just outside Geneva and completed in 1959. To appreciate the size of this accelerator, consider the scale of the plan (Fig. 9.16): the magnets weigh 3200 tons, and stand in a ring that is 200 metres in diameter. To produce particles at 28 GeV, the proton synchrotron uses 5 or 6 megawatts of power. It cost $30m to build, and many millions of dollars annually to run. Here we have Big Science on the scale of a large steel works or motor car assembly plant.

In this pursuit of the infinite, however, enough can never be enough. The largest accelerator now in service (1973) is the *Fermilab* machine near Chicago. According to the original design, there were to be 1000 magnets around the circumference of a circle more than a mile across, set

9.16 CERN site layout.

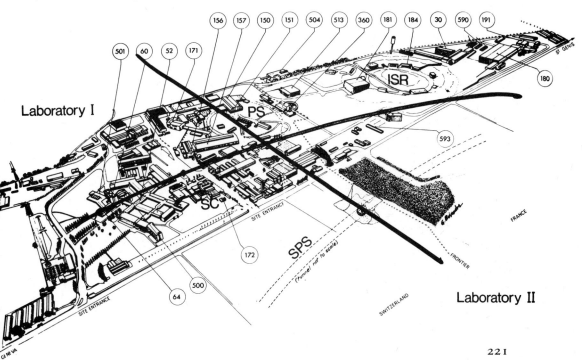

in a great tunnel (Fig. 9.17) under the earth to shield the harmful radiation. In fact this plan was just too expensive; instead of producing 500 GeV particles as originally intended, the initial phase was reduced from $400m to $250m, with a beam of only 200 GeV.

To appreciate the overall magnitude of high energy physics, we need to look not only at the largest machine of each epoch, but also at the number of particle accelerators that now exist in the world (Fig. 9.18). These are to be found mainly in the United States, Europe and the Soviet Union – but smaller instruments, that would have been regarded as enormous only a generation ago, are installed in universities or government research establishments in other countries.

BIG SCIENCE FACILITIES

The sheer cost of research in high energy physics will be discussed in the next chapter. But there are other consequences which are no less important.

We note, for example, that the task of designing and constructing a

9.17 The lowering of the first magnet into the main ring of the Fermilab Accelerator (June 1970); 1000 such magnets will eventually fill the 6.4 km circumference of the ring.

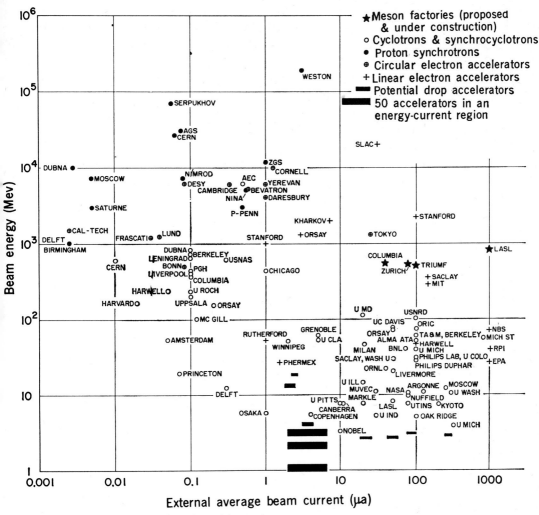

9.18 Worldwide inventory of particle accelerators.

'research facility' on this scale can no longer be left to the scientists themselves. It has to be done by engineers and other technical experts with special experience in this particular type of work. A subsidiary technology of 'accelerator design' has thus grown up, with a turnover comparable with that of a minor industry such as the manufacture of motor cycles. But this technology is not productive in itself: it only exists to serve the purposes of the research scientists who plan the experiments for which the instrument was built.

Merely to keep such a machine running, to control all the auxiliary equipment, to carry out routine maintenance and repairs requires a large staff of engineers and technicians, working in three or four shifts to

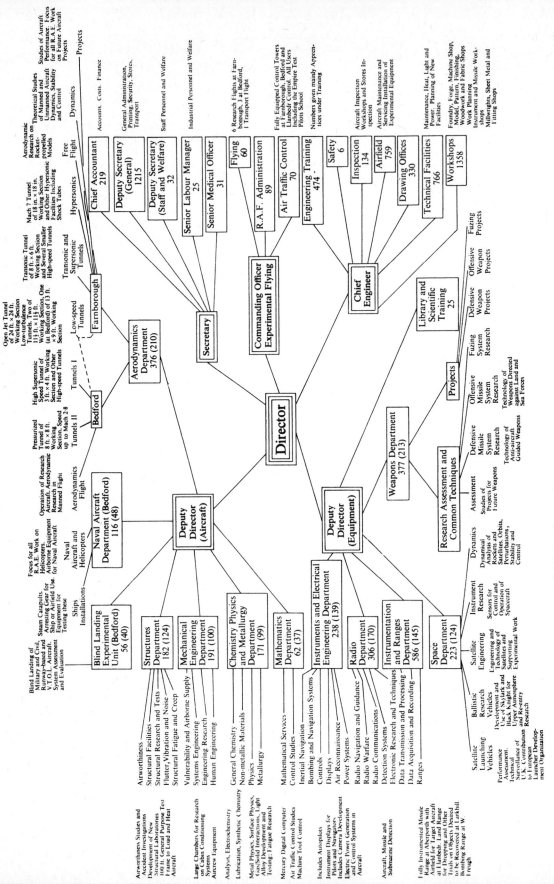

9.19 Diagram of the organization of R.A.E. Farnborough.

ensure continuous working. These too are highly skilled specialists, not doing research themselves, but quite essential to the overall scientific enterprise.

The enormous cost of each machine demands very large 'utilization factors'. The design must allow for a number of experiments to be carried out simultaneously, with very rigorous timetables and efficient use of every opportunity. The research programme must be adapted to the facilities available, and the research workers must be disciplined to work within these constraints.

Big machines lead inevitably to Big Science – that is to large-scale research organizations, with planned programmes. A 'laboratory' such as CERN is a small city of several thousand people gathered on a single site. In addition to the accelerators and their ancillary equipment, there must be auxiliary services such as canteens, a library, stores, lecture rooms, reception offices, a telephone exchange etc. This assembly of people and things must be rationally ordered: it becomes a *bureaucratic* organization, with a Director, Assistant Directors, Departmental Managers, personnel officers, purchasing clerks, accountants, contract draftsmen and so on (Fig. 9.19).

The management of such an organization is not the sort of task for which an old-fashioned professor is normally trained. As we have seen, scientific authority in the intellectual sphere does not necessarily carry with it the ability to handle men and money on this scale. If the administrative jobs are to be done by people with genuine scientific experience – and there are very good reasons why the responsible policy decisions should not be put into the hands of lay 'managers' – then means must be found to choose and train a number of scientists in the special skills of institutional diplomacy, committee craft, public relations, budgetary control etc. A new career structure, a whole new profession, of 'science management' or 'research administration' is growing up in and around Big Science.

The 'sophistication factor' in scientific apparatus applies not only to such large machines as particle accelerators, reactors, or space-rocket systems: in the effort to squeeze the very last ounce of observational accuracy from an experiment, the research scientist strives continually to refine his techniques. If a 50,000 volt electron microscope gives valuable information about the structure of living tissue, then a 100,000 volt

9.20 Victorian spectroscopist.

machine will give better information. Why not build a microscope at 250,000 volts or 500,000 volts or a million volts? A great deal of the active work in the physical sciences is devoted to instrument development, so that almost every simple piece of handy apparatus (Fig. 9.20) becomes a new complex, delicate and expensive device, which has to be designed, manufactured, paid for and maintained before it can be used for research. Whether or not some of these improvements (Fig. 9.21) are mere commercial gimmicks, this type of equipment is now taken for granted in all scientific research laboratories. The minimum stock of such apparatus, the building to house it, the technicians to run it, the budget to pay for it can only be found if many users band together. 'Little Science', in the sense of research work by individuals or small groups, can only be done within a Big Science organization such as Bell Labs (see p. 199) that provides these facilities (Fig. 9.22). The commercial pressures to sell us these facilities parallel the political pressures of the space race or the accelerator game of leap frog.

Another source of Big Science is the big mission. As we shall see, one

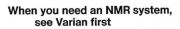
9.21 Contemporary advertisements for scientific apparatus.

9.22 Bell Telephone Laboratories at Murray Hill.

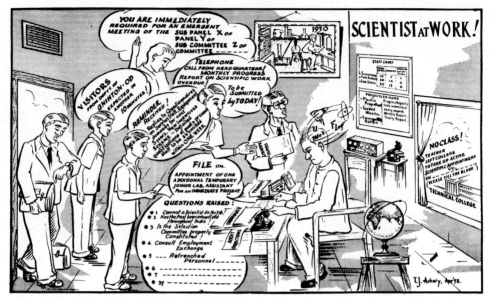

9.23 Cartoon kindly supplied by Dr D. S. Kothari for whom it was drawn when he became Defence Research Adviser to The Government of India.

of the lessons of the Second World War was the power of a large research and development organization actually to achieve almost any conceivable technical goal. If $1000m is being spent on the design of a supersonic transport aircraft, or on an anti-ballistic missile system (chapter 13), or even on a cure for cancer, then it is spent in an organized, planned, bureaucratic manner employing many thousands of 'qualified scientists and engineers', in large establishments with carefully designated objectives.

A very high proportion of the world's scientists and technologists are thus to be found within the chunky institutions of Big Science. For the leaders, the burdens of administrative authority cannot be shrugged off. This is the modern scientist's dilemma: should he refuse managerial responsibility, and thus remain only an individual research worker without the power of a big team at his command: or should he accept this responsibility and become so loaded with administrative duties that he has no time for genuine scientific work (Fig. 9.23). The temptations of bureaucratic office are great, because these are the posts which are officially honoured: but the loss of personal intellectual authority in the eyes of the scientific community comes hard if this is the source of one's self-esteem as a scholar.

228

From the worm's eye view of the typical research worker, the main feature of Big Science is that he must work in a *team*. The investigation itself is on a far larger scale than his personal responsibility, initiative or control. He no longer *does research*: he *takes part in a research project*.

Team research in science dates back to the nineteenth century. The German professor with his assistants (p. 59) would be working together within the same field, cooperating in the attack on the same basic problem. In so far as the assistants were indebted to the professor for ideas on which to work and technical advice as they proceeded, they might be considered a loosely knit team acting as an extension of the intellect of the leader. In the British tradition, the Cavendish Laboratory under J. J. THOMSON and RUTHERFORD was much more than an association of independent individuals who happened to be studying the atom or the nucleus (Fig. 9.24).

9.24 Research students at the Cavendish Laboratory, June 1898. From left to right; back row, S. W. Richardson, J. Henry; middle row, E. B. H. Wade, G. A. Shakespear, C. T. R. Wilson, E. Rutherford, W. Craig-Henderson, J. H. Vincent, G. B. Bryan; front row, J. C. McClelland, C. Child, P. Langevin, Professor J. J. Thomson, J. Zeleny, R. S. Willows, H. A. Wilson, J. Townsend.

PS COUNTER EXPERIMENTS APPROVED BY THE NPRC

Table 1E

EXPERIMENTS FINISHED DURING THE PS YEAR (1.1.72 -23.12.72)

Expt. Code	Code	Beam Description	Description of Experiment	Authors	Date Approval/Completion	Total weeks	Status
SIC5	m_9	Enriched K, $\bar{p}$ < 4.1 GeV/c	Polarisation in backward scattering. $\pi^- p_\uparrow \to p\pi^-$, $K^+ p_\uparrow \to pK^-$, $\pi^+ p_\uparrow \to \Sigma^+ K^+$. Polarised target, scintillators, wire spark chambers. HP 2116 B	CERN-Trieste Collaboration: Bradamante,Conetti,Daum,Fidecaro G., Fidecaro M., Giorgi, Kalmus G., Piemontese,Penzo,Schiavon,Vascotto	6. 5.70 / 28.10.70 / 3.11.71 / 31. 5.72	2 (T) / 10 / 7 / 5	Submitted for publication
P7	k_{17}	Separated K, $\bar{p}$ < 1 GeV/c	Studies of $\bar{p}$, K^- and Σ^- atoms. Stopped beam. X-ray detection with Ge(Li) solid state detectors	Karlsruhe-Stockholm Collaboration: Backenstoss,Bergström,Buracciu, Egger,Hagelberg,Hassler,Koch,Povel, Rolli,Schwitter,Tauscher	3.11.71 / 5. 7.72	2 Periods / 2 "	Beam Tests finished
S 91	d_{30}	Unseparated $\pi^\pm$, $K^\pm$, $\bar{p}$ < 12 GeV/c	K^+p, K^-p, $\bar{p}p$ forward and backward scattering, annihilation of $\bar{p}p$ in 2π's or 2 K's at small angles (high energy part). C magnet, on-line wire chambers, gas Cerenkov counter, IBM 1800	CERN-Ec.Pol.,Paris-Orsay (Acc.Lin.)-Stockholm Coll.: Baglin,Briandet, Carlson,D'Almagne,Damerell,Eida, Fleury,Gracco,Homer,Johansson, Lehmann P., Lundby,Navarro,Pevsner, Ratcliff,Richard,Rosny,Treille,Tso	8. 4.70 / 28.10.70 / 1.12.71 / 31.5. 72	4 (T) / 6 / 5 / 2	Partly published
SIO4	d_{30a}	Unseparated $\pi^\pm$, $K^\pm$, $\bar{p}$ < 12 GeV/c	Strangeness + 1 missing mass in $\pi^- p \to \Lambda^0 + M$. Scintillators, spark chamber, water Cerenkov counter	University of Rome-RHEL Coll.: Dore,Guidoni,Laakso,Marini,Martelotti,Massa,Piredda,Pistilli,Conforto, Hart,Mallary,Middlemas,Rosner, Walker	6. 5.70 / 4. 6.71 / 2. 2.72 / 31.5.72 / 30. 8.72	2 (T) / 2 / 4 / 3	Analysis
S 99	m_{11}	Low-energy separated beam to produce high flux of $\bar{p}$ between 0.6 and 2.0 GeV/c (Modified q_9)	Differential cross sections for $\bar{p}p \to \bar{p}p$, $\pi^-\pi^+$, K^+K^- between 0.6 and 2.0 GeV/c. Wire chambers, counters, AEG magnet	QMC-RHEL-DNPL-Liverpool Coll.: Kalmus,Gibson,Eisenhandler,Hojvat, Williams,Lee Chi Kwong,Usher, Pritchard; Astbury,Jones,Arnison, Parsons;Kemp,Woulds,Range,Harrison	5.11.69 / 23.10.70 / 12.10.71 / 1. 3.72	5 (T) / 6 / 6 / 3	Analysis
S 93	b_{19}	Short neutral beam derived from eg with a vertical septum	$\Phi^\pm$ measurement by time dependence of $K^0 \to \pi^+\pi^-$ and of the charge asymmetry in leptonic decay. Charpak chambers,wide gap magnet, large H_2 Cerenkov	CERN-Heidelberg Collaboration: Eisele,Filthuth,Geweniger,Gjesdal, Luth,Kamae,Kleinknecht,Presser, Steffen,Steinberger,Vannucci,Wahl	4. 6.69 / 28.10.70 / 4. 6.71 / 12.10.71 / 30. 8.72 / End Oct. 1972	6 / 11 / 8 / 3	Partly published

9.25 Committee Report, CERN.

Nevertheless, even under the most inspiring leadership, any competent research worker amongst the assistants would be expected to work steadily on his own, and to be treated as if he were free to do his own experiments and to publish his own papers. The custom by which the head of the laboratory puts his own name as a co-author on to all such publications has always been considered a mild abuse of authority, and a symptom of decadence in those institutions where it is practised. Whatever the realities of power, patronage, and prerogative, the myth of personal independence is the official norm in academic research.

In Big Science all this is changed. It is clearly understood that each research 'project' is a collaborative effort, where the individual is personally responsible only for one technical aspect of the work. Let us see what this means even in the pure and basic science of high energy physics.

The first stage is a 'proposal' (Fig. 9.25) made by a research group to the high and mighty committee that controls access to the accelerator beam and to the ancillary facilities. Since each experiment may cost millions of dollars, and there is much competition between the various groups, this is perfectly reasonable. But it does mean that decisions must be taken, from above, concerning the likely outcome of each proposal and the competence of each team in carrying out its plan. In a multi-national institution such as CERN, or even in a national laboratory such as the Rutherford Laboratory in England or Brookhaven in the United States, such decisions may force groups from different universities to collaborate as a single team, with consequent problems of joint leadership and organization.

The second stage is setting up the apparatus (Fig. 9.26). This means

9.26 Layout of low momentum K⁻ beam for the 1.5 m hydrogen bubble chamber at the Rutherford Laboratory.

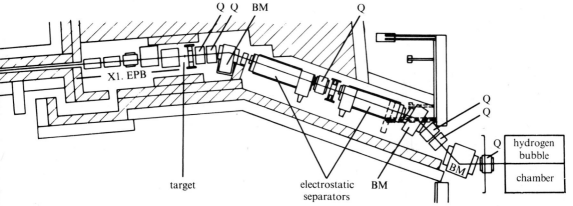

9.27 Apparatus used in experiment on antiproton–proton elastic scattering and two-body annihilation.

careful design or lining up of very complex equipment – enormous magnets that will draw off and focus a beam of the required particles, various arrays of counters, spark chambers, bubble chambers etc., all linked automatically by electronic circuits and controlled by computers.

The individual components are themselves very large and complex devices (Fig. 9.27) which cannot just be taken down from the shelf or ordered from the laboratory workshop, so that careful coordination with the technical staff of the laboratory will be needed. Remember that this is very precise engineering, working at the limits of technical expertise at each stage, not mere routine assembly of standard items. Testing the equipment and finding faults may involve many months of effort before it is in proper working order.

The third stage is the actual running of the experiment. Because high energy particles can only be produced by random processes, such as the bombardment of a block of material by the energetic protons in the primary beam of the accelerator, it may take weeks or months of continuous running (Fig. 9.28) to accumulate sufficient data to prove the point at issue. This means three-shift working, with long nights of attention by the members of the team to make sure that their apparatus is properly adjusted and that they are not wasting the time allotted to them on the big machine (prime cost – say £20 per minute)!

When the experiment is finished, they must still analyse the 'data'. This may consist of hundreds of thousands of bubble chamber

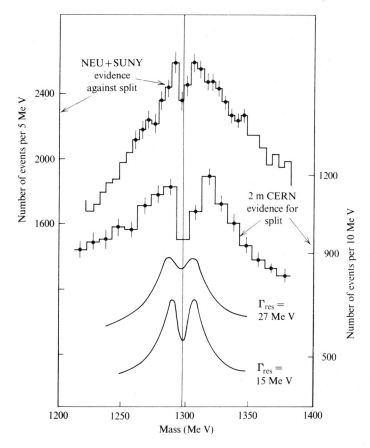

9.28 Statistical uncertainty in experimental results in high energy physics; evidence for and against the 'A$_2$' meson. Diagram from A$_2$ debate letter by Bogden Maglich.

9.29 Eighteen-prong event produced by a 16 GeV/c π^- beam in the 2 m CERN hydrogen bubble chamber.

photographs (Fig. 9.29), showing the tracks of various particles taking part in various interaction or collisions. These have to be scanned in the search for the rare 'events' for which the experiment was planned – a highly skilled and very exacting task. This enormous labour has to some extent been automated by special microscopes connected to a computer; but the analysis of the experimental results in high energy physics (Fig. 9.30) remains a major task in itself and cannot be done by one man with a slide rule plotting points on a grubby piece of graph paper.

The ALVAREZ group at Berkeley (Table 9.1) is probably the largest high energy physics team in the world – but its size is not untypical, and would be completely outclassed by many groups in space research. The point is not so much that a large number of technical activities and duties need to be coordinated and planned, but that all these people are working on a single scientific experiment. This means that the graduate students and young Ph.D.s – supposedly research workers in the making – are forced to become highly specialized in their functions. One man becomes expert in electronics, another handles the computer programming, a third is really occupied in managing the group of ladies who do the scanning, and so on. *The boundary between science and technology has worn very thin.* The electronics man in this team is just as much an

9.30 An isometric view of the differential cross-sections in K^+p elastic scattering measured in Experiment 8.

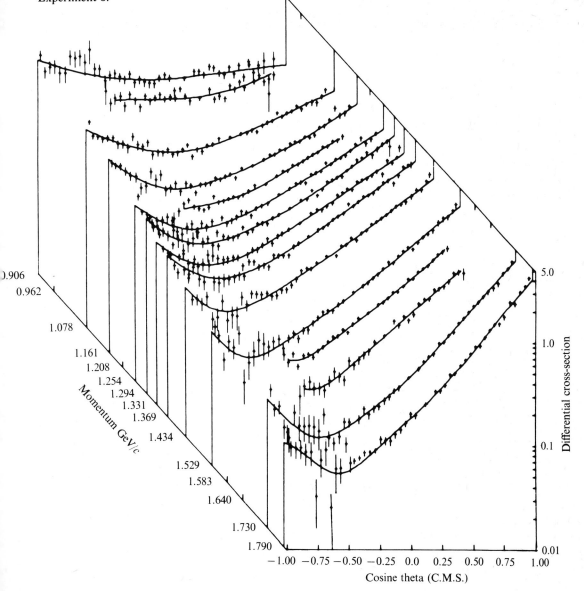

OBSERVATION OF A HYPERON WITH STRANGENESS MINUS THREE*

V. E. Barnes, P. L. Connolly, D. J. Crennell, B. B. Culwick, W. C. Delaney,
W. B. Fowler, P. E. Hagerty,† E. L. Hart, N. Horwitz,† P. V. C. Hough, J. E. Jensen,
J. K. Kopp, K. W. Lai, J. Leitner,† J. L. Lloyd, G. W. London,‡ T. W. Morris, Y. Oren,
R. B. Palmer, A. G. Prodell, D. Radojičić, D. C. Rahm, C. R. Richardson, N. P. Samios,
J. R. Sanford, R. P. Shutt, J. R. Smith, D. L. Stonehill, R. C. Strand, A. M. Thorndike,
M. S. Webster, W. J. Willis, and S. S. Yamamoto

Brookhaven National Laboratory, Upton, New York
(Received 11 February 1964)

It has been pointed out[1] that among the multitude of resonances which have been discovered recently, the $N_{3/2}$*(1238), Y_1*(1385), and $\Xi_{1/2}$*(1532) can be arranged as a decuplet with one member still missing. Figure 1 illustrates the position of the nine known resonant states and the postulated tenth particle plotted as a function of mass and the third component of isotopic spin. As can be seen from Fig. 1, this particle (which we call Ω^-, following Gell-Mann[1]) is predicted to be a negatively charged isotopic singlet with strangeness minus three.[2] The spin and parity should be the same as those of the $N_{3/2}$*, namely, $3/2^+$. The 10-dimensional representation of the group SU_3 can be identified with just such a decuplet. Consequently, the existence of the Ω^- has been cited as a crucial test of the theory of unitary symmetry of strong interactions.[3,4] The mass is predicted[5] by the Gell-Mann−Okubo mass

length of ~10^6 feet. These pictures have been partially analyzed to search for the more characteristic decay modes of the Ω^-.

The event in question is shown in Fig. 2, and the pertinent measured quantities are given in Table I. Our interpretation of this event is

$$
\begin{aligned}
K^- + p \to \Omega^- &+ K^+ + K^0 \\
&\downarrow\ \Xi^0 + \pi^- \\
&\qquad \downarrow\ \Lambda^0 + \pi^0 \\
&\qquad\qquad \downarrow\ \gamma_1 + \gamma_2 \\
&\qquad\qquad\qquad \downarrow\ e^+ + e^- \\
&\qquad\qquad \downarrow\ e^+ + e^- \\
&\qquad\qquad\qquad\qquad \downarrow\ \pi^- + p.
\end{aligned}
\tag{1}
$$

From the momentum and gap length measurements, track 2 is identified as a K^+. (A bub-

9.31   The names of the discoverers of the Omega-Minus particle.

engineer practising his craft as he would be if he were wiring up a power plant or a telephone exchange. Research itself – the tussle with a problem of natural philosophy – has given way to professional expertise in a variety of techniques.

The final stage of research is the publication of a paper. In high energy physics this document may bear the names of dozens of 'authors' (Fig. 9.31). This makes a mockery of the established procedures for the primary publication of the results of research. Such a paper can no longer be attributed to each of its fifty-seven authors as if it were his personal work. The internal equilibrium of the scientific community is threatened because judgements of originality, technical skill and other scientific qualities can no longer be made on the objective basis of published work.

This phenomenon is most notable in high energy physics, which is the most heavily 'industrialized' branch of pure science. But the same tendencies are to be found in all fields. Even in chemistry, where the typical research project is often the work of one man with the aid of many sophisticated instruments with highly skilled operators, the same trend of multiple authorship is to be observed (Fig. 9.32).

9.32 Incidence of multiple authorship in chemistry, as a function of date.

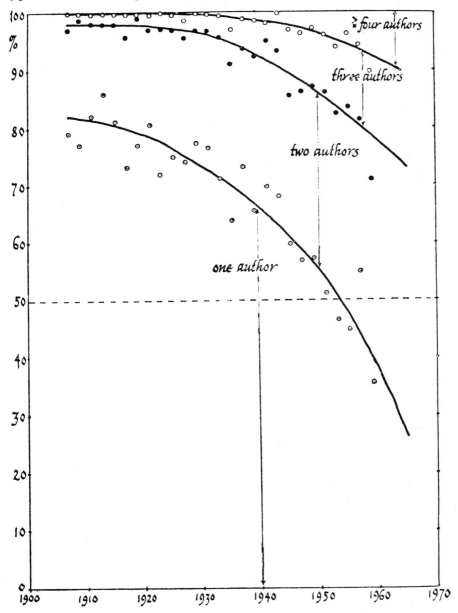

Team research in Big Science raises many new issues of practice and principle. The investment of money, and of human effort and loyalties in such corporate enterprises can heighten conflicts over scientific issues; or it can lead to a form of scientific log-rolling whereby merited criticism is suppressed out of respect for the livelihood of so many excellent people. Team science may be necessary for technical reasons, but it could be much less efficient intellectually than individual research.

The pressures of time and scale, the need for careful planning and control, make this sort of life no different from development work in industry. The spirit of personal achievement, or of independent artistry, is submerged in favour of group dynamics and disciplined craftsmanship. Perhaps, also, much genuine originality of mind will be suppressed as the organizers and impresarios take charge.

Who can adequately manage such groups? Their work becomes more and more subject to the 'authorities', who cannot really out-think the teams and the results they will obtain, and yet must somehow guess the outcome and decide which experiment is most likely to succeed. The present generation of research leaders were themselves trained in the older style of personal research with modest resources: their successors will not have had this experience, and must somehow demonstrate their scientific powers in this much more bureaucratic and industrialized environment.

What is the meaning of 'training for research' under such circumstances? How can a member of such a team internalize those standards of imagination and criticism, the balance between radical conjectures and conservative scepticism, which are the hallmark of the scientific mind?

To the extent that a team is self-governing and impersonal, it lacks ethical standards of behaviour, and may be driven by the institutional desire to survive at any cost. A large laboratory, whether organized around a particular big machine or merely created by administrative fiat, is potentially immortal. It can continue indefinitely by drawing in new people to replace the old. The natural selection processes of age and mortality no longer act against out-of-date problems. There is enormous human resistance to any administrative action to push such an institution along new roads, whether for the benefit of pure science or of human welfare.

In later chapters we shall consider the overall problems of the planning and financing of science, with its vast budgets and manpower surveys. What I have been trying to show here is that science itself, as a way of life, has changed enormously in the past half century, and is much less readily distinguished from ordinary industrial, commercial, managerial or bureaucratic existence than it once was. Paradoxically, the scientist who multiplies his observational or computational power manyfold with new and complex apparatus has become a slave to his own machines and to those who cooperate with him in their use. What he has gained in technique, he may well have lost in intellectual grasp and in the joys of the hunt.

10 PAYING FOR SCIENCE

In our day, the mania for planned, controlled, contrived, subsidised experiment, which, like war, appeals to the masses because of its exorbitant cost and its unanimous call for many democratic hands, is a syndrome of man unhinged in more ways than one.

C. Truesdell

THE COST OF RESEARCH

The cost of scientific research is not easy to grasp. Despite many published budgets from which various figures can be drawn, it is difficult to understand just where the money goes, and almost impossible to assess the value of the resulting products. What does research cost – and what is it worth to those who buy it? Science does not fit neatly into the economic categories of the commercial world.

What does it cost to do an 'experiment'? 'Cost' has no precise meaning

Table 10.1. *Physical sciences expenditure, Bristol University (1969–70)*

	£
Salaries of teaching and research staff	341,000
Other departmental salaries and wages	216,000
Departmental and laboratory costs (materials, equipment, etc.)	123,000
Total from main university funds	680,000
+Expenditure from research grants etc.	202,000
Total	882,000
No. of permanent staff, 80	
Cost per staff member, £10,000	

in this context, so let us consider a typical academic scientist teaching in a modern university and publishing, say, two or three scientific papers a year. He might be a faculty member of the Department of Physics or the School of Chemistry at Bristol University; in this group of laboratories he would have about 80 permanent colleagues, with 120 technical assistants and about 150 graduate students. Between them, they would be spending something like £1m per year (Table 10.1) of which 30–40% might be academic salaries and another 5–10% would really be spent on facilities for undergraduate teaching. (No attempt has been made in this chapter to quote the very latest figures. What with changes of economic policy, inflation, variations of exchange rates etc., financial data of this kind can only be kept up to date by continued reference to government reports etc. year by year. The reader should search out and compile the relevant information for himself. This is the only way of getting a feel for the quantities involved.) Thus, the research work of our Dr X would be costing more than £5000 a year in technical and secretarial assistance, experimental equipment and materials, laboratory and library facilities etc. In the United States the figure would be somewhat higher: a few years ago it was estimated that these costs for a full-time research worker at Ph.D. level in industrial research would be about $40,000–50,000 per year (Fig. 10.1). These costs are evidently rising rapidly, as more

10.1 Annual cost per scientist or engineer to large US industries that are significantly engaged in research and development.

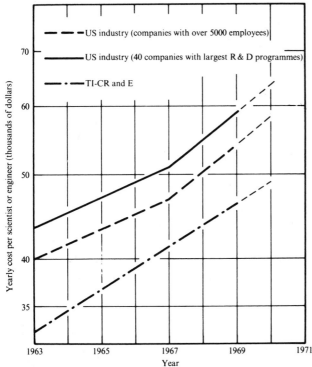

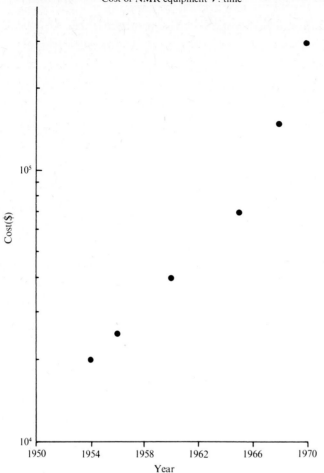

10.2 Increases in the cost of nuclear magnetic resonance equipment, 1950–70.

and more sophisticated and expensive apparatus is brought into use (Fig. 10.2).

In ordinary basic 'little' science, therefore, such expenditures, amounting to rather more than his personal salary, could not be met from the pocket of the individual scientist. In some special fields, by an extraordinary intellectual effort, good research can still be done with the proverbial 'sealing wax and string', but this is not reasonable for the million or so 'scientists' who now flourish in the world. Amateur research is now economically out of the question. There is no alternative to professional employment as a scientist by an organization that is willing to provide the equipment and other facilities that have become essential.

But a single research scientist on his own is no match for a team of Ph.D.s. Laboratory services and equipment are hideously expensive unless they are shared by several research groups. The minimum investment for a useful research organization in pure or applied science is thus of the order of £100,000 a year. There is no substitute for the massive funding of science by large-scale industry or by government: the financial resources of individuals or of small enterprises are quite inadequate even for 'little' science.

Big Science costs about 200 times as much. A big accelerator laboratory such as CERN (p. 221) spends about £30m a year (Fig. 10.3) to provide facilities for about 650 man years of research – that is, about £50,000 for each fully qualified Ph.D. This represents something like thirty distinct 'experiments'. A scientific paper in high energy physics is the final product of £1m worth of machinery, manpower and materials.

10.3   CERN budget.

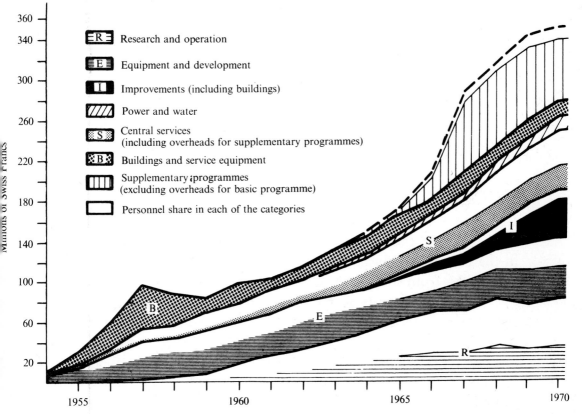

Table 10.2. *NASA R & D appropriation in millions of dollars (excluding heavy equipment facilities)*

Year	Total credits	Fundamental research	Applied research	Development
1960	363	97	166	100
1966	5071	598	713	3760

This branch of pure science has become so expensive that it can only be undertaken with the resources of a whole continent. CERN is remarkable as an extremely successful international organization; the research done there would be beyond the means of any single European nation.

Space research is bigger than ordinary Big Science, by another factor of 100. Back in 1967 the total budget of the US National Aeronautics and Space Administration (NASA) was £2000m just for 'experiments', not counting the cost of the rockets and other basic facilities (Table 10.2). Figures like this are incomprehensible except by comparison with the total national income of a country of several millions of people or the commercial operations of an international oil company. Research on this scale can scarcely be afforded by the world as a whole. Collaboration by the US and the USSR in the exploration of space is more than a gesture of friendship and reconciliation: it is an economic necessity if the work is to continue at all.

GOVERNMENT SUPPORT FOR SCIENCE

Let us return from the stratosphere to normal science at the university level. Where does the money actually come from to pay, say, for research in the physical sciences at Bristol University? Although it all comes, eventually, from the Government, the main channel still goes through the University Grants Committee and the general accounts of the University. The expenditure per student in faculties of science and technology is three or four times the expenditure in faculties of arts, law etc. This covers much more than the cost of the 'practical' laboratories. Research by university teachers of science is supported directly by the university because it is reckoned to be an essential adjunct to their educational duties. By providing these facilities and by rewarding success in research with promotion, the university fosters science without imposing a precise obligation to undertake any particular investigation. Within this institutional context, therefore, a university lecturer or professor can indulge in research almost as if it were a personal hobby, and is not bound professionally or contractually to produce scientific papers to order.

In practice, however, the cost of scientific and technological research has risen far beyond the normal income of even the richest universities. In all economically advanced countries, it is customary for university laboratories to receive additional grants of money for equipment, assistants etc. from outside sources – government departments, industrial corporations, private foundations, or international agencies. Typical of such grants are those received by the University of Bristol from the UK Science Research Council (SRC) which covers a range of subjects broadly under the heading of 'Physical Sciences and Engineering' (Table 10.3). Something like 25% of the total research budget of our scientific laboratories would normally come from agencies of this kind.

Table 10.3. *SRC grants current at Bristol, on 31 March 1967*

Subject	Value (£)	Ph.D. studentships
Astronomy	57,000	—
Biology	127,536	30
Chemistry	26,615	57
Mathematics	—	9
Metallurgy and materials	71,221	—
Nuclear physics	8,760	32
Other physics	35,449	
Space	3,275	—
Aeronautical and civil engineering	6,015	8
Electrical and systems engineering	15,150	4
Mechanical and production engineering	7,070	2
Total current	358,091	142
(Actual expenditure in year)	101,232	

In many other countries – especially the United States – research expenditure, as a whole, is not a larger fraction of the total university budget than in Britain, but the proportion of this supported from outside grants is much higher.

THE RESEARCH GRANT SYSTEM

How is this subsidy for research actually arranged? This is so much part of the life of the modern 'academic' scientist that it is worth description and comment. The British procedure, which is fairly typical of most Western countries, is as follows:

(i) Dr X, a lecturer at the University of Q, makes an application for a grant to undertake a special research – e.g. 'Investigation by atomic electric resonance of the W spectrum of Quodlibetium.' He explains why he thinks this would be scientifically valuable, and how he plans to do it. He needs £15,000 for a new atomic electric resonance spectrometer, and £2000 for each of the next three years for a research assistant. With miscellaneous further expenses, such as travel to conferences and computing, his request totals, say, £22,000.

(ii) The permanent officials of the SRC (or, as it might be in the United States, the National Science Foundation (NSF)) send copies of the application to 'referees'. These are experts in this branch of science: in practice they could only be Professor Y and Dr Z, who also do atomic electric resonance spectronomy at other universities, and who are in the usual scientific relationships of competitive collaboration with Dr X. Their task is not easy: if the proposal is truly interesting and original, there might be considerable uncertainty as to whether it will work; their opinion may well be based more upon a judgement of Dr X's reputation and professional skill than upon the merits of the scheme as such.

(iii) The referees' reports are considered by SRC or NSF officials, and eventually presented for decision to a 'subject committee' consisting of more scientific experts – for example, a committee of professors of atomic chemistry drawn from a number of universities. Their job is to choose between various similar proposals, on the basis of their scientific merits, and somehow to keep everybody reasonably satisfied, within the funds made available to them from some higher committee. This general procedure is called *peer group assessment*.

(iv) Dr X gets his grant – perhaps without the money for the research assistant, because funds are short – and then has considerable freedom to spend it as he thinks best within the general terms of his grant. Strictly speaking, the grant is given to the University of Q, but in practice the 'principal investigator' can regard it as his own research fund, and deals directly with the SRC over details.

(v) Three years, and several reports, later, the process is repeated: Dr X – now a professor – has a research group in full swing, and has become a pillar of atomic chemistry and an ornament to his university.

The great advantage of this procedure is that considerable financial resources can be swung into the support of new scientific work of 'timeliness and promise'. A bright idea can be exploited rapidly, or a new field opened up by systematic subsidy. Powerful new instruments can be placed in strategic centres of research, where they can transform techniques and methods.

Support for able young scientists, regardless of their position within the academic hierarchy, is advantageous. In the past, the tyranny of a powerful professor has often inhibited the imagination and originality of his subordinates: given the independent means of his own research grant, a junior faculty member can go ahead with his own research projects and snap his fingers at his reactionary or jealous elders.

But there are real dangers in the system. One form of corruption is the art of 'grantmanship' – the collection of a number of grants from outside bodies to support pretentious, poorly conceived research on a large scale. Independence of local control encourages the fragmentation of university departments, which tend to become mere federations of disjoint research groups, without coordination or plan. The department or faculty loses control over the people engaged in teaching and research – especially when the salaries of many members of the research staff are paid directly from research grants. Many technical problems of financial accountancy also arise, such as the commitment of 'university' funds to overheads and capital expenditure associated with research activities. These dangers have been most noticeable in the USA, where the cornucopeia of grants and research contract has flowed so much more generously; but they are not unknown in more sober nations.

The question one might ask is whether it is wise to put the control of all these funds in the hands of central groups of scientific 'peers'. At best this can lead to 'log-rolling', where Professors X, Y and Z make sure that each gets his share out of the pork barrel. At worst there are temptations towards genuine corruption or to the suppression of awkward customers whose research might invalidate the labours of the members of the Establishment and put them out of business. Multiplicity of sources of research funds, to match the critical and competitive character of science, may be much wiser in practice than tidy schemes of coordination and control.

Another task of the SRC, again through its expert committees, is to allot studentships to various universities for the training of M.Sc. and Ph.D. students. This would amount to something like another £100,000 per year at Bristol alone. Since most other sources of funds for this purpose are now insignificant, this is an important task. In effect, here is the valve controlling entry into the scientific profession and moderating the growth of the scientific community. The balance between different subjects and between their pure and applied aspects, over the next generation, depends upon the policy adopted at this stage. It is fairly certain, for example, that if ten times as many students are given grants for training in the theory of elementary particles as for theoretical chemistry, then there will be a great many more professors of elementary particle theory than of chemical theory in twenty years time.

In the UK, grants for training in research are still regarded as 'scholarships' for the benefit of the individual student: in most other countries, support for graduate students comes in the form of research assistantships, attached to the research itself. In other words, a research grant may provide funds for a certain number of graduate assistants, who take their Ph.D.s whilst working on the project of the principal investigator. Not only does this increase the fragmentation of the academic institutions, and the patronage power of the grantman: it is also another symptom of the complete professionalization of science, where the research student is no more than a paid apprentice to a particular speciality.

Look now at the over-all responsibilities of an agency such as the Science Research Council (Table 10.4). It not only supports university research in many small fragments of a few tens of thousands of pounds; about three-quarters of its expenditure goes in the running of big laboratories for full-time specialized research. This is the channel for the British subscription to CERN, and the source of funds for the big British accelerators at the Rutherford Laboratory and at Daresbury.

Notice, indeed, that nuclear physics takes half of this budget, and space research and astronomy another quarter. This is the meaning of Big Science in financial terms: the enormous demands of expensive equipment can distort the allocation of funds a very long way from proportionality to the social value of the research or to the numbers of scientists who will benefit by doing it. Yet the SRC is a body of

Table 10.4. *SRC expenditure for year ended 31 March 1968*

	£	£
London office administration	1,000,000	
Rutherford Laboratory (high energy physics)	7,000,000	
Daresbury (high energy physics)	3,200,000	
CERN subscription	5,000,000	
University grants for nuclear physics	1,270,000	
		17,470,000
ESRO subscription	4,500,000	
Greenwich and other observatories	1,130,000	
Radio and space research station	1,350,000	
Space research central expenses	1,000,000	
Astronomy central expenses	70,000	
University grants for upper atmosphere	570,000	
		8,620,000
Atlas Laboratory Computing	800,000	
University grants for other researches	6,570,000	
Post-graduate training awards	4,810,000	
NATO scientific schemes	320,000	
		12,500,000
Total		38,600,000

independent scientists representing a wide range of disciplines and is by no means under the thumb of power-hungry nuclear physicists. The economic logic of expensive instrumentation and of big team research creates its own scale of priorities and values within the scientific community itself.

The Science Research Council is only concerned with a part of the natural sciences. Research relevant to medicine, to agriculture, or to the natural environment (e.g. geology and ecology) is supported by three other research councils, whose total funds about equal those of the SRC (Table 10.5). The fruit research at Long Ashton, for example, is financed and administered by the Agricultural Research Council, although this laboratory is attached to Bristol University for academic activities. The total expenditure of the British government on relatively basic research is thus of the order of £100m a year. But notice that the cost of nuclear/high energy physics is still a considerable fraction of this sum, being comparable with the whole expenditure of the Natural Environment Research Council. The sum going to the Social Science Research Council looks much too small; is this due to a prejudice against academic sociology and economics, or is there a lot of money going into this sort of research from other sources? (See chapter 12.)

But the British government has other channels into which money flows for research and development. The total expenditure on *civil* research (Table 10.6) is about three times the amount spent in universities and by the Research Councils, and this total is nearly matched by defence research (see chapter 13). At this stage, however, we are including the costs of technological *development*, which is much more expensive than

Table 10.5. *Research Council expenditure (1971–2)*

	£m
Agricultural Research Council	18.7
Medical Research Council	23.0
Natural Environment Research Council	15.9
Science Research Council	55.7
Social Science Research Council	4.1
	117.4

laboratory experiments (see chapter 8). An engineering prototype, or pilot plant, can cost many millions, and demands enormous technical resources and skilled manpower. A conventional principle is that relative costs in basic research: development: production plant rise as 1 : 10 : 100. Nobody can prove that these scale factors are 'correct', but these rates give a rough idea of what must happen. If it is argued that only 'basic' science is truly 'science' – the rest being 'technology' – then the total science budget of £500m a year is not so generous after all.

It is interesting to see that the much larger R & D expenditure of the US federal government follows a somewhat similar pattern (Fig. 10.4). It would take a great deal of social analysis to work out the equivalences between agencies such as NASA or AEC and the corresponding British government departments, but the distribution of funds is broadly the same. Notice, however, that defence research in the USA (Fig. 10.5: 'External Challenge') approaches 60% of the total, whereas the British proportion is about 40%; perhaps this simply means that the richer

Table 10.6. *Central government expenditure on scientific research and development (1968–9)*

	£m	£m
Industry		
aerospace	102.8	
atomic energy	47.4	
general	19.0	
		169.2
Agriculture, fisheries and forestry	8.3	
Roads and transport	3.7	
Employment, industry and trade	0.6	
Housing and environment	0.6	
Law and order	0.6	
Health and welfare	2.8	
Universities	52.4	
Research Councils	76.7	
Miscellaneous (administration etc.)	13.3	
Total civil	328.2	
Total defence	237.2	
Total	565.4	

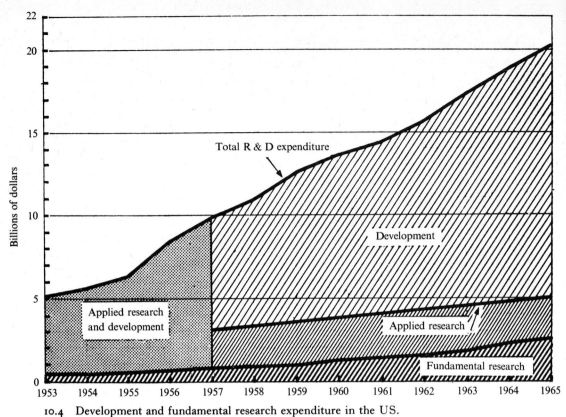

10.4 Development and fundamental research expenditure in the US.

10.5 R & D expenditure in the USA.

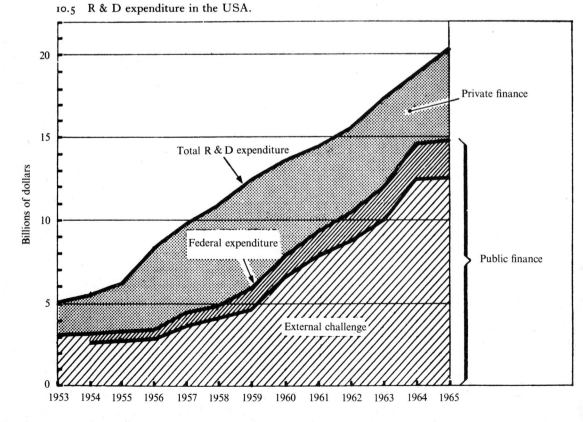

country can afford to waste even more of its income on such luxuries! As a small compensation, let us remember that the US Department of Defence was in the curious habit of supporting a great deal of purely academic research at universities, and played rich uncle to some very elegant but useless scholarly ventures. Applied research and development are recorded as nine-tenths of the total but this may have been achieved by application of the rule of thumb rather than by strict accountancy.

One big difference between British and American methods of financing research is that the US government farms out about 60% of its research work in the form of 'contracts' to private industry (Table 10.7).

Table 10.7. *Central government spending (1967)*

	USA		UK	
	$ billion	%	£ million	%
Intramural	3.4	21.3	238	44
Industry	9.7	60.1	199	36
Universities	2.1	13.2	62	11.3
Others	0.8		46	
Total	16.0		545	

In Britain, there is a greater tendency to create special government laboratories – for example, the Royal Aircraft Establishment at Farnborough and the Royal Radar Establishment at Malvern – for this type of work. The American method has its benefits in the cost-plus profits of the industrial corporations, and also in their attitude to research and innovation. Notice, however, that the proportion of government money for science flowing into the universities is much the same in both countries: it is just the absolute amount that is so much larger in the United States.

Research and development expenditure by private corporations in the UK and in the USA is about equal to government expenditure in the respective countries. The accountancy is a bit complicated by the nationalized industries but the general pattern is again much the same

(Tables 10.8, 10.9). The most striking feature of these figures is their uneven distribution. Scientific industry is heavily concentrated in a few industries, such as aerospace and electrical engineering, whilst it is almost negligible in others of comparable economic and social importance. How could one justify spending 100 times as much on aircraft (no, this is not all for defence!) as on ships, or railways? It would be interesting to know whether R & D expenditure in the Soviet Union is similarly unbalanced, but the real facts there are carefully hidden in budgets, plans and manpower statistics.

GROWING EXPENDITURES

It is obvious that a gross total sum of money for 'national R & D expenditure' includes a diversity of disparate items, from the academic salary of an Einstein to the capital cost of a rocket test rig. But the growth of this sum over the past few decades is a major economic phenomenon.

Table 10.8. *R & D spending in UK industries (1967–8)*

	£m
Aerospace	146.6
Electrical engineering	142.7
Mechanical engineering (very miscellaneous!)	56.5
Motor vehicles	33.6
Chemicals and coal	39.8
Metals and metal products	23.9
Pharmaceuticals and plastics	23.4
Food, drink and tobacco	15.4
Petroleum and rubber products	14.7
Scientific instruments	12.7
Textiles, man-made fibres and clothing	11.9
Stone and clay products	8.5
Timber, furniture, paper, printing, etc.	8.2
Ships	2.3
Railways	1.7
Construction	4.4
Other	7.0
	553.3

The data for the USA (Fig. 10.6) have been conveniently represented on a logarithmic scale: in the twenty years from 1945 to 1965 R & D expenditure has multiplied by a factor of fifteen. In that same period, the national income (Gross National Product, GNP) has multiplied by three. In other words, R & D expenditure has risen from about 0.6% of GNP to about 3%, where it is tending to level off. This is the financial counterpart of the continuous growth of scientific activities of all kinds, such as published papers (p. 56), over the centuries, with an extra cost factor for the increasing sophistication of experimental equipment during the past few decades. But notice that educational expenditure in the USA has multiplied by ten since the end of the war, so that perhaps the research component of education has not grown so rapidly after all.

Similar historical trends are to be observed in all advanced countries (Fig. 10.7). In 1934 J. D. BERNAL complained of the meagre £6.6m being spent on research in Britain; compare this with the present total, which exceeds £100m. Generally speaking, a modern industrial nation in Europe (and, of course, also Japan) spends 1–2% of its GNP on R & D

Table 10.9. *Research and development as percentage of net output, 1958*[1]

	US companies	UK companies
Aircraft	30.9	35.1
Electronics	22.4	12.3
Other electrical	16.3	5.6
Vehicles	10.2	1.4
Instruments	9.9	6.0
Chemicals	6.9	4.5
Machinery	6.3	2.3
Rubber	2.7	2.1
Non-ferrous metals	2.0	2.3
Metal products	1.3	0.8
Stone, clay and glass	1.2	0.6
Paper	0.9	0.8
Ferrous metals	0.8	0.5
Food	0.5	0.3
Lumber and furniture	0.2	0.1
Textiles and apparel	0.2	0.3
All industries	5.7	3.1

[1] From B. R. Williams, *Minerva* **3**, 61.

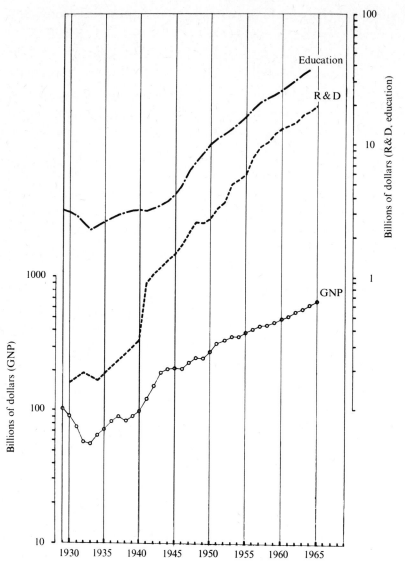

10.6   Comparative growth of GNP, expenditure on education and on R & D in the USA.

and about one person in a thousand of the population is a 'QSE' (Qualified Scientist or Engineer). These figures already represent quite a heavy drain on resources of money, skilled technicians, sophisticated equipment, educational facilities, and, above all, intellectually gifted and well-motivated people. It is hardly likely that they could be further multiplied by a factor of five or ten over the next few decades. Indeed the

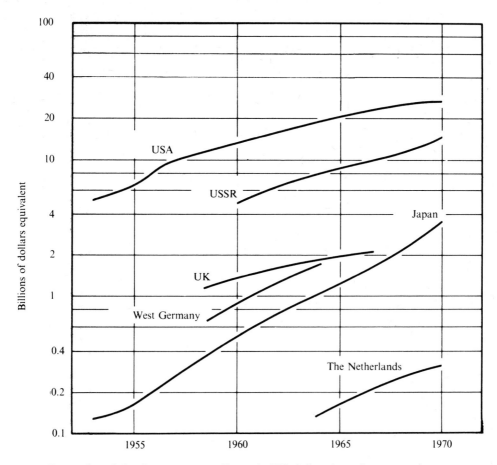

10.7   Research and development expenditures in US dollars in various countries.

fact that almost all these 'science indicators' have ceased to grow since the early seventies is scarcely surprising. The saturation point had to come soon. Science may not yet have reached its full stature as a social force, but the problem now is to decide relative priorities rather than to find yet more men and money for research of all kinds.

Since scientific research is a 'luxury' – or at least a marginal provision for a distant future – it is not surprising that a rich country spends a higher proportion of its national income on R & D than a poor country. There is a rough correlation between the ratio R & D/GNP and *per capita* income, country by country round the world (Fig. 10.8). Some politicians attempt to turn this correlation on its head: by spending an inordinate amount on science, a government can pretend that it is richer

257

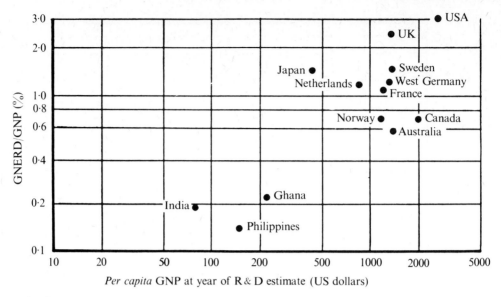

10.8   Gross national expenditure on R & D as percentage of GNP.

than it really is. In other words, a national research organization becomes a 'status symbol' in the public image of that country, entitling it to entry into the scientific club (Fig. 10.9). The reality is that the total research effort of a country where the people are very poor can never be very large: the contributions of the 'developing countries' to world expenditure on science are practically negligible, although the research that they do may have a significant effect on local conditions in each region (see chapter 11).

THE ECONOMIC RETURN ON RESEARCH

Why should such large amounts of money be spent on science? What is the return on the investment? This is a question that cannot be answered solely in economic terms. Science has its spiritual and aesthetic values, which cannot be translated into pounds or dollars. We might, of course, play the parlour game of counting each published scientific paper as a unit of 'scientific value produced'. On this basis Derek de Solla Price has estimated the 'size' of science, country by country, and shown that there is a typical establishment of about one scientific author for every $10m of GNP (Fig. 10.10). Notice that India fits this line quite well, whereas the

ATIONS CAN PUBLISH OR PERISH

Derek J. de Solla Price

**Toting up science papers is an absurd device
for evaluating the productivity of researchers,
but it is a surprisingly useful approach
for comparing the science programs of nations**

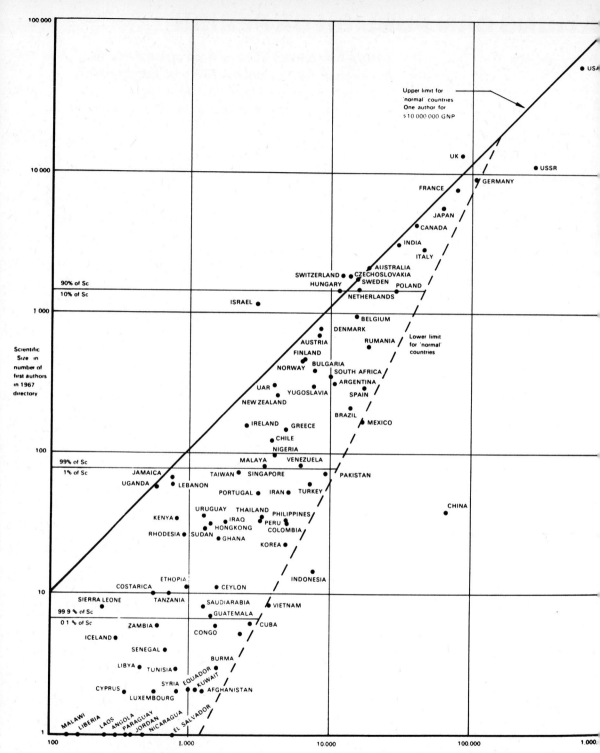

10.10　Number of science authors as a function of GNP, showing that 90% of scientists are in the few richest countries, etc.

UK and Israel are apparently spending too much on science. On the other hand, the Latin American countries 'could do better'. But this is misleading: merely counting quantities of papers tells us nothing of their *quality*. Factors of 10 or 100 in the real value of each unit contribution would completely outweigh the counting of names in an index of authors. A minister of finance would be very foolish if he assumed that his country was scientifically advanced and competent simply because it had the requisite ratio of professors and Ph.D.s.

It was for a time argued that research and development effort had a direct effect on industrial growth and therefore, simply by voting larger sums to science a country would inevitably make itself richer within a few years. The causal connection is obvious enough in principle, but it has not turned out that way during the recent past (Fig. 10.11). The UK and the USA spent three times as much, in proportion, as West Germany and Japan, yet grew economically at less than half the rate. Australia manages to grow richer on a very modest scientific expenditure – probably

10.11   The relation between R & D expenditure and growth for various countries.

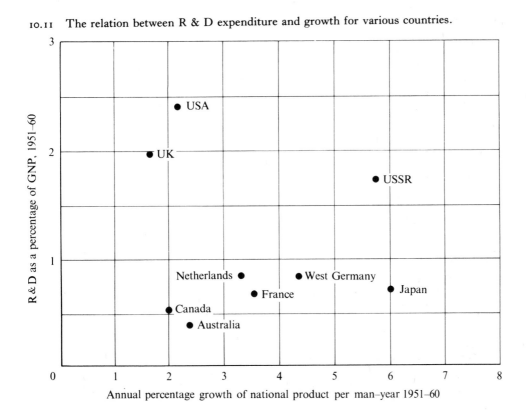

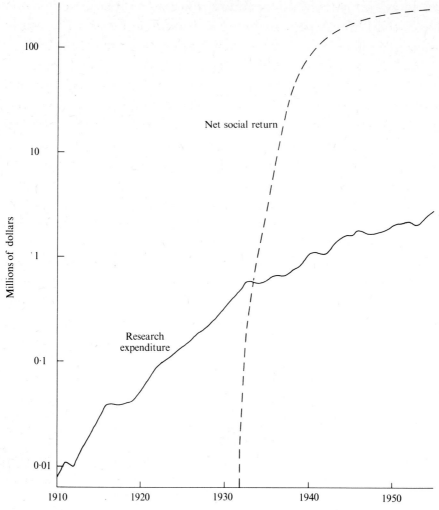

10.12 Annual expenditures, and calculated economic benefits from research on hybrid corn.

because advanced techniques are imported there by the big international corporations who do their research work in the metropolitan countries (p. 269). The important factor is not the absolute amount spent on science, but the right mixture of basic research and development applied to the available economic resources of the country.

In a phase of disillusionment about the magical powers of scientific expenditure, it is well to emphasize that the long-term profitability of research, in purely economic terms, is not in doubt. The calculation is not easy, but a famous and successful example of directed research has been analysed in detail (Fig. 10.12). From 1910 until 1933, sums totalling about $3.5m were spent by various agricultural research organizations in the USA on the development of a commercial variety of hybrid corn (p. 34). In 1933 the first reliable variety was supplied to farmers, and gave an increased crop yield of 15–20%. The extra corn thus produced without additional effort has been worth about $200m per year ever since. The formal calculation of the 'rate of net social return on the investment' can be varied according to the economical hypotheses one starts with but whatever way one looks at it one gets a profit of about 700% per annum.

But even in this obvious case of highly successful research leading to an immensely profitable innovation, the response of the farmers was not instantaneous (Fig. 10.13). It took about twelve years for hybrid corn to take over the whole corn belt of the USA. For each individual farmer the calculation of his own advantage in making the change could be quite complex, and might depend sensitively on the scale of his operations. And although the cost of this research now looks very small beer, it was beyond the resources of ordinary seed merchants and plant breeders.

10.13   The spread of hybrid corn: areas that planted 10% or more of their corn acreage to hybrid seed in specified years.

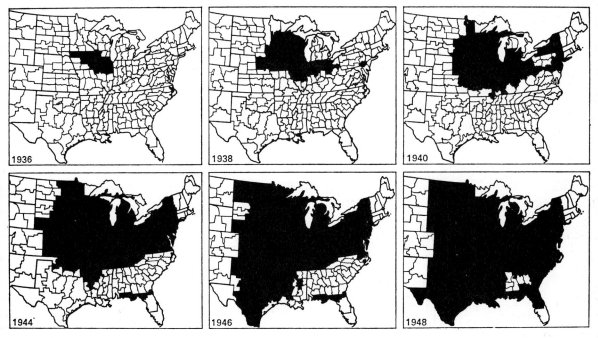

Only the deliberate and sustained efforts of permanent organizations such as Federal and State agricultural research stations, over a period of twenty-five years, without certainty of a final return, achieved the result.

It would also be wrong to deduce that all research expenditure, if continued for sufficient time, must eventually pay off at this enormous rate. In this case, for example, we should include all the money and effort that must have been spent, from the beginning of the century, on unsuccessful proposals for increasing the corn crop of the USA. For every research project that breaks through at last, there must be five or ten that end nowhere, or only add to basic knowledge, or are called off for accidental reasons before coming to fruition. Research is much more of a lottery than an investment: the pay-off can be very large, but it is uncertain and may come only after a life-time of effort.

Even in a situation where the outcome is reasonably predictable, as in the development of an existing device according to conventional technical principles, substantial economic resources are needed. Consider the following hypothetical case, discussed by Lord BLACKETT. A manufacturing firm decides to develop a new line in, say, oscilloscopes. This would take several years for a team of, say, ten QSEs. Allowing laboratory costs etc. of about £10,000 a year for each member of the team, the company would be facing R & D expenditure of about £100,000 a year. But few firms can set aside more than, say, 10% of their turnover for R & D, so they would be expecting sales of the order of £1m a year. If each oscilloscope costs £1000, this means selling at least 1000 instruments a year, which is perhaps the size of the whole British market for such a specialized device. Unless it had a very large export business, therefore, this firm would need to have a near monopoly position locally. But with only a single product it would be very vulnerable to technical competition. In other words, this sort of development project would be very risky unless the company were much larger, selling many different instruments to the tune of £10m a year or more, and employing several thousand people. If we were talking of computers, or power plants, with each item costing hundreds of thousands of pounds, then these figures would need to be multiplied by factors of 10 or 100.

This is the fundamental reason why research is mainly financed by very large corporations or governments which have the means to surmount the threshold for successful development by big teams of scientists and engineers. The fundamental ideas may come from low cost individuals, but large-scale technology demands risk capital in very large amounts.

11 SCIENCE AS A CULTURAL IMPORT

He died in exile; like all men, he was given bad times in which to live.

Jorge Luis Borges

From Imperial China to Inca Peru, every civilized society had the makings of what we now call science. Technical inventiveness, as in Chinese porcelain; mathematical dexterity, as in the Mayan calendar; philosophical acuity, as in the Hindu sacred writings; curiosity concerning natural phenomena, as in all human societies: these ingredients of the scientific attitude are readily found in any culture that has left a record of its thoughts and activities. The achievements of an ancient civilization are an excellent source of pride for any nation in the process of regeneration. To discover and admire these achievements is one of the main tasks of the historian and archaeologist.

But there is no real evidence that anything like the scientific revolution of seventeenth-century Europe had taken place, or was about to take place, in any other region of the world. Whether by accident or by historical inevitability, the ingredients were there mixed in the right proportions, and subjected to the necessary religious temperatures and political pressures, to react and combine into a new cultural compound. Modern science, everywhere in the world, has grown from this single source. Along with the steam locomotive, the machine gun and the Christian Bible, science was exported from Europe in the nineteenth and twentieth centuries, and imported into other countries as a minor component of Western civilization. In the era of colonialism and imperialism, there was no attempt to graft Western scientific methods and ideas on to the stock of philosophy and technique already existing in countries such as India or Japan: the complete system, as currently expounded and practised in Cambridge, Paris and Berlin, was to be accepted as a whole. The Second Law of Thermodynamics is an integral part of the steam-engine package.

In this chapter, we shall take a look at the world-wide diffusion of science that has taken place during the past century. Unfortunately, this subject is not well-documented, either historically or in its current state. There seems to have been no serious scholarly effort to describe this diffusion as a social phenomenon in its own right. 'Science' has been supposed vaguely to go along with 'technology' and 'industrialization', as a small but essential factor in 'modernization'.

SCIENCE AND SOCIAL DEVELOPMENT

Even in an 'advanced' country such as Britain the relationship between scientific research and technical innovation is not at all simple: in a 'developing' country, such as Brazil, where almost all the techniques of industrialization have been imported as blueprints or machinery, the causal connections become even more tenuous and obscure. The vast literature on the ideal organization for government research on agriculture, or on the cost-benefits of a nuclear/fertilizer/desalination/irrigation complex tells us almost nothing about the capabilities and morale of the small groups of individuals who must man these desirable developments. It is just taken for granted that educational institutions copying Harvard or MIT, backed up with a few automatic spectrophotometers and research reactors, will deliver the goods. Disappointment at the small returns from such expensive investments, and the resulting disillusion with science as the catalyst for social development, is now a significant political factor in the Third World: failure to think carefully about the rationale of science in modern society is now having serious consequences where it hurts most.

AUSTRALIA

Consider the case of *Australia*. By the beginning of the twentieth century this country, despite its recent colonial origins, had several large cities where people lived at the peak of current technical style. Universities modelled on those of England and Scotland had been founded in the 1850s and contained flourishing schools of science, engineering, medicine, etc. There were local learned societies: the Royal Society of New South Wales was founded in 1821. In the 1920s a rather good government organization (CSIRO) was set up for agricultural research.

But until after the Second World War, in the physical sciences Australia was a desert. Industrial research did not exist, and academic science was of very low quality. Holland, or Denmark, carried more weight in the scientific world. Why was this so? Partly because Australia lived mainly on its agricultural and mining products, and imported most manufactured goods. There was almost no local market for the knowledge that might be produced by physical scientists, and no material incentive for giving them financial support. The non-material incentive of national pride in scientific achievement was also inactive. The universities had no graduate schools: an able student automatically set out for England, to take a Ph.D. at Cambridge and to make a scientific career in Britain. Or he would come back as a lecturer to an Australian university, and quietly settle down, discouraged from research by meagre resources and local incomprehension of its purpose. With universal literacy, and close cultural ties with Britain and the USA, no Australian could say that he was denied direct access to the latest scientific knowledge: but geographical distance, economic circumstances, and the historical political relation with Britain kept that great continent in a state of deep cultural *provincialism*. Australian science was weak and fragmented because nobody believed in it as a local product: physics was what one did at the Cavendish Laboratory, or at Göttingen, not in Sydney.

Since the War all this has changed. Deliberate policies for support of university research, and for the creation of new government laboratories have been extremely successful. Many first-rate Australian scientists, who had lived or would have stayed in Britain, have returned home, to found genuine research schools. In radio astronomy, for example, Australia is a world leader, and in many other fields the work is of the highest international standard. One does not need to calculate some parameter such as the production of scientific papers per head to verify that Australian academic science has 'taken-off' in the sixties and is now as healthy as anywhere in the world.

This transformation was relatively easy because Australian urban life is essentially the British way of life transported abroad and climatically modified. Australian and British scientists can move readily from one habitat to the other without severe cultural shock. Given adequate facilities for research, and efficient communication channels with the rest of the world, most Australian scientists prefer to live and work in their own country. Entice a few experienced research workers home together, and an intellectually stimulating and self-sufficient community can soon be created. In the era of jet travel, science in Brisbane or Adelaide need be no more provincial than in Aberdeen or Aberystwyth.

But research in Australia is still very meagrely supported by private industry. Australian manufacturing industry is dominated by international corporations, such as General Motors, that rely upon the research laboratories in their home countries to provide fundamental technical knowhow and new scientific inventions. Australia thus enriches herself from the general pool of modern technology, but does not contribute to this in proportion to the wealth and education of her people. Applied science in Australia is mainly carried out in government laboratories and is still mostly concerned with agriculture and the use of primary products such as wool.

JAPAN

The great success story of industrial development is *Japan*. When it was reopened to foreign intercourse in the middle of the nineteenth century Japanese culture was at a mediaeval level. But from about 1870 onwards there was a deliberate, rational, government policy to import European techniques and knowledge. Newly founded colleges employed foreign teachers, and many Japanese were sent abroad to study. For each special technique, the most advanced country was chosen: Britain for machinery, geology and shipbuilding; France for biology and mathematics; Germany for physics, chemistry and medicine; the United States for industrial methods and agriculture. The rapid success of this policy, especially in military and industrial techniques, is a familiar fact of world history.

Until recently, however, Japanese research science did not match this material development. Engineers and technologists were trained primarily to use conventional European methods and designs. Success in steel-making, shipbuilding, aircraft manufacture and other advanced techniques was achieved by employing large numbers of carefully trained technical experts rather than by the invention of new methods. Research facilities were meagre, both in private industry and in government laboratories. The charge that Japan thrived commercially by imitating foreign industrial products is essentially just.

Despite its substantial scale, research in the universities was not really on a much firmer basis. By the 1930s, Japan was a major contributor to the world total of scientific literature, but university research was very poorly supported and much of the work was of low quality. The cultural bridge from Germany to Japan was not sufficiently broad, and the attempt to transplant Western science into a completely different intellectual environment had been too hurried. The Japanese student of physics who spent a few years working for his Ph.D. in Berlin learnt to carry out the formal manipulations of experiment and theory, but with an inadequate grasp of the language and this very brief experience of the research atmosphere he could not be brought to think for himself as an independent scientist. To put it crudely, academic science in Japan was largely imitative and lacked the creative spirit of genuine natural philosophy. It is interesting to note, as a measure of the place of science in Japanese life, that little use was made of the intellectual resources of her large scientific community during the Second World War. There were no Japanese scientific weapons to compare with the atomic bomb, radar, rockets, jet aircraft, or penicillin: the university laboratories, instead of being used for research, were gradually transformed into specialized munitions factories.

Since the War, however, Japanese science has grown to genuine maturity. The switch to the manufacture of sophisticated electronic and optical equipment demanded much heavier expenditure on research and development. The research laboratories of the big industrial corporations are comparable in scale and quality with those of their American and European competitors. The very large numbers of scientists and technologists employed in Japanese industry are well educated and very expert. With the excellent apparatus that they now manufacture for themselves, they can do research as well as anyone in the world.

Yet Japanese industrial R & D is still dominated by the old policy of looking out for other people's new ideas, and putting them into production. It is still true to say that Japan makes better American electronic equipment than the Americans, better German cameras than the Germans, better Italian cars than the Italians, and almost as good Swiss watches as the Swiss. Despite such remarkable achievements as the 'bullet' railway system and the supertanker, Japan still has an adverse balance of trade in patent royalties. If industrial innovation depends in the last resort on the technical originality and the creative imagination of the scientist or inventor, then these abilities are not yet adequately encouraged in Japanese life.

Academic science also is now at a good international level of quality. Just how this was achieved is not quite clear. After the War, Japanese science was in complete disarray and for a long time was very isolated from the world scientific community. Perhaps this period taught hard lessons of self-criticism and independence of mind, and evolved a new generation of leaders who had fully internalized the scientific attitude. It was not simply a matter of better material resources. The laboratories in the major universities are as well equipped as is necessary for the research that they do, but financial support for apparatus is still inadequate by American standards. Japan is weak in the 'Big' sciences, such as space research, and has, of course, kept out of nuclear weaponry and similar extravagances; but in many less spectacular fields her scientists are amongst the world leaders.

Nevertheless, it is worth remarking that the characteristic qualities of Japanese research are those of the careful observer, the skilful craftsman, the diligent technician and the well-informed critic. Japanese academic life, protected by geographical and language barriers from the infective informality and spontaneity of the American intellectual community, evidently gives too little weight to the artistic, intuitive, speculative style in research. But this, too, may be an historical consequence of the painstaking German academic pattern on which Japanese science was originally modelled; in any case, high standards of craftsmanship and criticism are not so bad a basis for future scientific achievement.

INDIA

British officials with amateur interests in natural history brought European science to *India*. To satisfy various civil and military needs, the British government established geodesic, geological and botanical surveys, medical and health services, and schools of engineering. But there was no attempt at all at research that might be of industrial value, and until the 1920s only a smattering of science was taught in schools and universities.

This is scarcely surprising if one considers the official attitude towards science education in Britain at that time. The ideal of the well-rounded arts graduate, schooled in the classics or history, had been exported to India in the nineteenth century, and remained in fashion there long after it was seen to be out of date in England itself. Some people argue that this peculiar form of education was deliberately imposed on India in order to produce docile government clerks; whether or not this is true, it has dominated the Indian universities ever since.

Nevertheless, in the 1920s there was a spectacular flowering of native Indian scientific talent. The leading figure was c. v. RAMAN (1888–1970), a fascinating personality in the old style of the 'man of science', interested in all sorts of natural phenomena, an amateur whose obsession with research carried him from a comfortable civil service career into a poorly paid professorship at Calcutta, and eventually to international fame and the Nobel Prize for Physics. By some historical accident, Calcutta University became the well-spring for other distinguished physicists such as s. n. BOSE (1894–) and m. n. SAHA (1893–1956). Most members of this group did spend short periods abroad, but it is fair to say that their basic scientific background, in education and research, was Indian. Looking at Indian academic science at that period, one might quite reasonably have regarded Calcutta, Madras, Lahore and Bangalore as embryo Harvards and Göttingens for an immediate renaissance of Indian scientific culture.

India has now a large scientific community, but this expectation has not been fully satisfied. In the 1930s, physics everywhere was still in the 'sealing wax and string' era (p. 215). Apparatus was simple, cheap, and largely home-made. The material poverty of the Indian universities was not such an obstacle to good fundamental research. It was also a very personal activity, in the style of the private scholar, without the necessity of team work and corporate action. Indian science in the Raman era seems to have lacked any cohesion beyond the sphere of each professor with his assistants and pupils. Harsh competition for academic appointments and autocratic behaviour by some of the Grander Chams inhibited the growth of a true spirit of community. Old-fashioned in style, and without an industrial basis to give it meaning, academic science in India was at the mercy of quarrelsome individuals and external economic circumstances. The expansion of scientific activity since national independence was not well-ordered: resources are distributed very unevenly and scientific standards vary more than in any other major nation.

At the top we find a few élite institutions of matchless quality in staff, equipment and research output. The best-known example is the Tata Institute for Fundamental Research in Bombay. This was founded in 1945 by Homi BHABHA (1909–66) who had returned to India after a brilliant decade in Cambridge as a theoretical physicist. Connections with the millionaire Tata family, and close personal relations with Nehru, gave him the money and authority to build a 'Centre of Excellence' for basic physics in India. In scientific terms, the Tata Institute has been entirely successful and a proper object for pride. Yet the emphasis there on the most irrelevant branch of pure science is not healthy in a country that is so weak in bread-and-butter research. There is a tendency for the staff to set themselves apart, and not to involve themselves in the general problems of science in their country. Characteristically, they show their lack of confidence in Indian science by publishing all their work in American, British, or other foreign journals.

On the next level we might place the Indian Institutes of Technology, founded in the sixties to provide sophisticated engineering training. Through their connections with particular advanced countries, these are reasonably well-equipped, and have been able to attract a number of Indian scientists who had been working for some years abroad. It is still uncertain whether the 'brain drain' was harmful in the long run, to Indian science or Indian society. The loss of skilled men can be very serious – but since the research jobs previously open to them in India were not properly backed up with material facilities it is not obvious that they would have done much more if they had stayed at home. It may be more significant that those who eventually return to India bring back with them the scientific standards, the research style and the technical methods of Berkeley, of Oxford, of Harwell, of Bell Labs, where they have held responsible posts. The fact that Indian scientists are fluent in English and share much of the basic culture of the English-speaking world, gives them much more immediate access to American and British science and society than their Japanese contemporaries. If Indian science in the twenties was patterned on Oxbridge and the Royal Society, in the sixties and seventies it is looking much more towards MIT and the US National Science Foundation for its models.

Research in Indian private industry is still quite dead. To fill the gap, the government set up a large number of new laboratories, for aeronautics, solid state physics, atomic energy etc. But these laboratories are of very variable quality. The experienced applied scientists to administer them were just not available. The university graduates who come in with Ph.D.s or M.Sc.s are themselves so scientifically and socially naive that only determined leadership can make them useful. With poor pay, inadequate facilities, and a sprawling bureaucracy, it is scarcely surprising that they have disappointed expectations. But here again, the injection of strong doses of foreign experience by scientists returned from abroad, and the pressures of economic and technical reality as India becomes more industrialized, are having their effect.

In the worst position are the universities and colleges. Burdened with masses of students, poor in staff, buildings and equipment, often disrupted by strikes and demoralized by the high level of graduate unemployment, they can scarcely be regarded as inspiring centres of high science. Apparatus for research – even for teaching – is entirely inadequate and antiquated. Indian university physicists concentrate on theoretical work because they have no apparatus to do experiments: but neither have they computers with which to evaluate their formulae. Scientific curricula are intensely formal, abstract and out-of-date; but the collegiate organization of the universities makes reform almost impossible.

Yet there remains a touching devotion to the ideal of the scholar dividing his time between teaching and research. Many young lecturers and readers have long lists of published papers, and there is fierce competition for academic preferment on the basis of published work. Much of this research is very academic and pedestrian, even when it is accepted by a British or American scientific journal; but the stamina and intellectual vigour of those who produce it is not to be doubted. In various places, in various universities, a number of excellent scientists are struggling for recognition. If these talented people were exposed to a somewhat broader view of their scientific specialities, and given the material resources of their European contemporaries, then India would quickly become a major power in basic science. It is the poverty of the nation as a whole that confines their creative abilities. If these abilities could be harnessed more directly for the enrichment of the people, then these bonds, too, would rapidly be loosened.

Indian society, at large, has almost no comprehension of the meaning and power of scientific and technological research. The Indian scientific community is too large and too fragmented to act as a significant force on its own. The range of abilities and skills that it covers is too wide, and the general attitude to science is too bookish and unrealistic. But there are now to be found in India many technically competent, well-motivated, critical and imaginative scientists, able to undertake or direct first-class research in all fields. To that extent the diffusion of modern science into India is an accomplished fact of history.

PROBLEMS OF SCIENCE IN THE THIRD WORLD

The growth and spread of science is evidently a complex process, with very different characteristics in different countries. The mere factor of industrial development is not a sufficient explanation: general cultural, economic and political forces are also significant. In the case of *Canada*, for example, the proximity of Big Brother across the border is the main factor. In *Brazil* and *Argentina*, we must take account of economic incompetence and political brutality in explaining the weakness of the scientific community. As for *China*, nobody knows yet whether the determination to put science to work for the people has really succeeded, or what has been the effect on research standards of two decades of isolation.

AN IMAGINARY COUNTRY WITH REAL PROBLEMS

The state of *Saturnia* is not, therefore, to be found on the map, although it has been a member of the United Nations since it achieved national independence. It can scarcely be described as a very small country, but its 20 million inhabitants are poor, largely illiterate, and mainly concerned with making the best they can of their individual lives. The modest institutions of higher education inherited from the past have been expanded into substantial universities, crowded with students seeking formal qualifications for permanent jobs as government officials or teachers. For most of them, therefore, it scarcely matters that the curriculum has not changed for half a century, nor that each subject is taught separately and dogmatically, as material to be learnt by heart and played back faithfully at examination times.

The number of students of pure science is, in fact, quite small – no more than enough to provide replacements for the professors who teach them. Indeed, it is surprising that there are any science students at all, since this is the worst-taught subject in the few secondary schools through whom the Saturnian intelligentsia all come. The career of the would-be university teacher is not very attractive at the outset. For ten years as a teaching assistant he must answer the beck and call of his professor, pass numerous examinations proving his detailed knowledge of all aspects of his subject that were once of scientific interest, and make sure that a sufficient number of publications have appeared with his name on them. But then, when he himself becomes a professor, he can look forward to many comfortable years on a modest middle-class salary, which can easily be supplemented by putting his name to this or that commercial advertisement or political document, thus maintaining the social standards of the academic profession without too much intellectual effort.

The main obstacle to automatic progress into this socially advantage-ous niche is the need to do something called 'research'. At this early stage in its history as an independent nation, Saturnia has no graduate schools awarding the Ph.D. It is necessary, therefore, to undertake advanced study abroad. Armed with a government scholarship, the brilliant young scientist takes himself off to the United States, or to the Soviet Union, or to the metropolitan country of colonial days, in whose language he has largely been educated. For three, four or five years he lives more or less as any other graduate student of some great university – Chicago, Moscow, Paris – as a member of the research team of an internationally famous professor, working with the most up-to-date and sophisticated apparatus upon the most fashionable problem at the very forefront of knowledge. His Ph.D. is hard-earned, but well justified: his name appears amongst the authors of several papers in the most reputable journals, and his dissertation shows that he has adequately served an apprenticeship in the crystallographic analysis of proteins, in the design of space rockets, or in mathematical logic.

Returning, thus intellectually and psychologically equipped, to his native country, he sets himself enthusiastically to further research in the same line. Alas, the resources of Saturnia are quite inadequate for such delicate tasks. Scientific apparatus is crude and obsolete. Laboratory technicians are no better than bottle-washers. Money for equipment and materials comes in derisory sums. Foreign exchange regulations delay for months the expenditure of a few dollars on minor spare parts for imported instruments. New books and journals arrive years late. Bureaucratic administrative procedures hamper every initiative and waste half his working day. Although the courses are out-moded in content and style, teaching duties are heavy, and less enterprising colleagues are jealous of his research achievements. Nepotism, political pressure, professional favouritism and internal university intrigues deny him the promotion he deserves, and sour his spirit. He is continually aware that his research interests have no roots in his own country; worst of all is the isolation from the competitive, critically exacting, but stimulating scientific community to which, for a while, he had belonged. Either he writes a few trivial papers, succumbs to local pressures, and takes with relish to the Saturnian pastime of academic Snakes and Ladders, or he writes in desperation to his former research supervisor and gets himself a scientific job in a technically advanced country, down the Brain Drain.

Yet he once had the talents to be of genuine benefit to his country. The natural resources of Saturnia, properly exploited, could make it rich enough. Money for economic development is not quite lacking, and many factories are producing goods for the local market. But these industries depend upon foreign capital and foreign technology. Complete factories, including office furniture, instruction leaflets and production engineers, are imported and set up as 'turn-key' projects, ready to turn out Saturnian radio sets, or Saturnian zip fasteners, at the touch of a button. Saturnian scientists are not employed to adapt existing techniques to local use, or to invent new products. Foreign experts supply the knowhow; the 'natives' merely contribute cheap labour and expensive, inefficient management.

Nor are these deficiencies being made up by government action. It would be altogether too tedious to recount the political saga of Saturnia during recent years. Demagogic politicians, financial adventurers and reactionary soldiers have all had their turn in power, and all shown their complete inability to grasp the place of science in this new nation. Police action has driven a number of the more outspoken academic scientists into exile, and keeps the remainder in sullen silence. Each successive government raises a new group of colonels or lawyers to administrative power, each with a different attitude to research.

Once upon a time, for example, there was a far-reaching plan to transform the nation by advanced science and technology. A network of research councils, research institutes, research bureaus and research laboratories was projected: directors were appointed, some buildings were started, and generous budgets for staff were allocated. Of this scheme, however, the only survivor is the *Saturnian Atomic Energy Commission*, whose chairman knew how to accommodate himself to the demands of successive governments. Its pride is a nuclear reactor, bought for several million dollars from an American firm, and now nearing completion. This device will not, of course, produce any saleable electric power, but is designed as a 'research facility'. Indeed, the SAEC employs two or three experimental physicists who are building apparatus that will use the modest beams of neutrons produced when the reactor eventually goes critical.

The other major research schemes initiated in that epoch have not proved so successful. The Saturnian Space Research Institute was disbanded abruptly before it could move into its new building, and the Cancer Research Programme had to be wound up when its director, the famous Saturnian Nobel Laureate, decided to return to his former Chair in the United States. But Saturnia boasts a National Chemical Laboratory where the effluents from the local brewery are analysed for arsenic, and the Government Meteorological Bureau is located near the International Airport.

To many far-sighted Saturnians, the most serious weakness in government research appears in the field of agriculture, by which, of course, 90% of the population make their living. The former colonial power did, indeed, set up a number of agricultural research stations and experimental farms, which supported the production of cash crops for export from large plantations. This research continues. But with many more mouths to feed, high-yielding varieties of food crops for local consumption are urgently needed. Despite considerable expenditure on agricultural research, Saturnia has made little progress in this direction. Indeed, the number of Ph.D.s actually doing research in agriculture is decreasing, since those few who chose to study this uninspiring science prefer desk jobs in the Ministry of Agriculture in the capital city.

In the last couple of years, however, there has been a great awakening in Saturnian government circles to the importance of *relevance* in scientific research. Environmental studies are to be emphasized, and only projects of technological significance are to be supported. It has been decreed, for example, that every university teacher is to undertake research on a relevant topic – either the detection and elimination of a dangerous pollutant, or the invention and manufacture of a technical, agricultural or domestic device. For each such project, a time limit of two years has been proposed. Meanwhile, 'academic' research will be forbidden in the university during official working hours.

This account of the state of science in Saturnia is necessarily somewhat sketchy. To complete the picture, one would need to compare and contrast it with conditions in the neighbouring countries of *Neptunia* and *Urania*. *Neptunia*, indeed, is so primitive that the recent creation of a university has been a real milestone along the road to progress: the first Neptunian Ph.D. will be returning there shortly to be Professor of Chemistry. The *People's Republic of Urania* on the other hand, in its dash along the road to industrialization, has rapidly expanded its scientific cadres. Whether, under the leadership of the People's Progressive Party and its heroic ally the Soviet Union, it will manage this subtle process more successfully than Saturnia, remains to be seen.

THE TRAGIC REALITY

The problems and difficulties that are suggested in this caricature of science in a developing country are not funny. For many highly intelligent, thoughtful, patriotic and sincere people, struggling for self-achievement and for the enlightenment of their country, these circumstances can be deeply tragic. European and American scientists have not always shown much sympathy for their position: they have blithely trained foreign students in the most sophisticated branches of pure research, without regard to their future circumstances, and then asserted arrogantly that of course nothing but practical engineering should be attempted in those countries. The machinery of international agencies such as UNESCO is too cumbersome, too closely tied to government circles, to deal with such delicate issues. The gossip journals of the scientific world such as *Nature* and *Science* say almost nothing about science in the Third World: not only is Saturnia a distant country, whose scientists count for little internationally; there is also a convention of courtesy and reticence so as not to make life yet more difficult for the Saturnian scientists in their struggle for survival against local enemies.

There is an urgent need for much more open discussion of these issues. The place of basic research in a developing country is not quite clear in principle, and may be very complicated in practice. The training of scientists, technologists, and technicians, is not simply a matter of creating a large number of universities or institutes of technology on the European pattern, and needs to be argued out anew. The relationship between applied science and industrial and agricultural development cannot be settled by a formula valid for all political and economic circumstances. If those who must take decisions on these matters are to act wisely, they must be fortified with knowledge of the experience of others in similar circumstances and with a stock of imaginative proposals drawn from a wide range of people and places. Only the truth will set them free.

12 THE SCIENCES OF SOCIETY

Whoever reflects on this cannot but marvel that the philosophers should have bent their energies to the study of the world of nature, which since God made it, he alone knows, and that they should have neglected the study of the world of nations, which since men made it, men could come to know.   *Vico*

Since science has an immense influence on society, why not solve all our problems with a science *of* society. Let us turn the instrument of scientific method on people and their social institutions: let us observe, analyse, predict and control human behaviour, just as we observe, analyse, predict and control the behaviour of electrons and amoebae. Such a science will, surely, be more influential and efficient as a social force than any mechanical contraption.

This is an ancient dream. Every social reformer believes, and claims, that he has the rational 'scientific' way to happiness or prosperity. PLATO's *Republic* exalted Philosophy. Thomas HOBBES (1588–1679) in *Leviathan: or the matter, form and power of a commonwealth ecclesiastical and civil* (published in 1651) made a scientific analysis of political behaviour to justify absolute autocracy. Two centuries later, Karl MARX (1818–83) claimed the intellectual authority of 'scientific socialism' for his predictions of revolution. The academic study of politics is often called 'political science' even by those who teach that it is an intuitive craft.

Another cliché is that 'the social sciences have yet to find their Newton', indicating that the whole subject is still in a very primitive and incoherent stage of development, without powerful unifying principles from which firm deductions can be drawn. The implication is, of course, that a 'Newton' will indeed eventually appear, and pronounce social laws by which we shall be able to chart political events with the same precision as we may predict the conjunctions of the planets, and sit down to design a new society as we might design a new aircraft.

Until that happy day, however, we must make do, as best we can, with what can be discovered by lesser men. There exists a vast body of learning, under the heading of 'the social and behavioural sciences', which shares many of the characteristics of the physical and biological sciences, and which is directly concerned with Man as a social being. It is impossible here, of course, to attempt to summarize the content of these disciplines – anthropology, sociology, social psychology, economics etc. – but our main theme would be incomplete if we did not try to say something about their general claims to scientific validity, and about their role in society itself.

GRAND THEORY

We might begin with a viewpoint that is nearer to mediaeval scholasticism than to modern science – the attempt to define a formal theoretical structure, within which all observable social phenomena may be supposed to be fitted. It is not easy to give a simple example of this type of sociology because the whole system is usually expounded in very abstract general language and at great length; any single paragraph would appear meaningless on its own. Precisely because human behaviour has such immense, subjective, emotive significance to the human observer, it is felt necessary to describe it in a neutral phraseology, defined by explicit axioms and formal deductions. Precisely because of the immense variety of incidental circumstances in actual social events, the style is dry and abstract, emphasizing the generality of the phenomena being discussed. No doubt the serious student gains some insight from mastering such a theoretical system; but since it scarcely says anything at all about any particular social situation, and almost every statement within the theory turns out to be true by definition, it cannot be verified or falsified by appeal to experience. Nobody with an ordinary training in scientific research need be deceived by an attempt to copy the style of axiomatic theoretical physics – which is a pretty sterile discipline, anyway!

Nevertheless, if any progress is to be made with research on more limited problems, it is necessary to have a general picture of the workings of human society. This picture is provided, in its details, by *political*, *social* and *economic historians*, and other academic experts. Just as we live on a particular planet, and in a universe whose features have been explored and mapped by geographers and astronomers, so we live in particular countries, in a world community whose social characteristics and institutions must be discovered, described, and recorded. Mere 'facts' do not make a very lively science, but they must be known and agreed if 'theory' is to have any basis in reality. The statement that the reign of Charles I was separated from that of Charles II by the government of Oliver Cromwell is as much a piece of scientific knowledge as that the Mediterranean is separated from the Atlantic by the Straits of Gibraltar, or that the head of a horse is connected to its torso by a neck. Whatever theories we may eventually erect to explain these facts – the facts will remain.

It is scarcely surprising, however, that historians, economists and political philosophers have attempted to arrange these facts into theoretical structures. History would be merely an endless succession of detective investigations if it were no more than a record of the 'crimes, follies and misfortunes of mankind'. Such theories, like those of the geologist or cosmologist interpreting the traces of very distant events in space or time, are likely to be highly speculative, but they provide a focus for further research.

Consider, for example, the writings of Herbert SPENCER (1820–1903), a dominant intellectual influence in Victorian England. A contemporary of DARWIN (p. 98), he developed his own ideas on the evolution of social organizations by analogy with biological organisms. In the era of Progress, he described the historical transformation of primitive communities into industrialized civilizations as a process akin to the origin and transformation of biological species, governed by the principles of natural selection and the survival of the fittest. We see now that this biological model of social change is much too simplified and naive, but it made some sense of a vast body of miscellaneous information about many different societies, and emphasized that every society, like every organism, is a structure of interdependent parts, each of which has a discernible function in the life of the whole. Then or now, such a simplistic description need no more be accepted as gospel truth than, say, the Big Bang theory of the origin of the Universe or any one of a dozen hypotheses concerning the origins of the solar system. Yet an apt metaphor may be more valuable as a guide to thought than a whole volume of 'data'.

ANTHROPOLOGY

In the late nineteenth century, attention began to be given to many aspects of human society that had been neglected by the historians, who could only study literate communities. A vast body of information was collected about myths, totem groupings, kinship systems, marriage customs, initiation rites etc. *Social anthropology* is a bit like natural history: whatever one would like to prove theoretically, the main objective is to build up a reliable record of the different ways in which people in different societies actually behave. Modern biologists affect to despise taxonomy; but without an adequate systematic knowledge of the different species that exist in nature, it would be very difficult to do reliable research in physiology or pathology. We now know that all human societies, whether savage or civilized, have well-defined and fairly permanent structural features, which can be compared and classified in various instructive ways. This again is a genuine scientific task requiring professional skill and expertise – especially when the information has to be gathered in an outlandish language, under uncomfortable physical circumstances.

It must be admitted, however, that this mass of information has not been reduced to order by any general theory that is universally accepted. Social and cultural anthropologists of different schools of thought emphasize different aspects of social behaviour, and direct their research towards the confirmation of a multiplicity of hypotheses and theoretical principles. But the data are extremely qualitative: it is very difficult to assign numerical values to the material of ritual and myth, kinship and child-rearing practices, and causal relationships cannot be checked by experiment. The problem of finding overwhelming evidence to prove or disprove any conjecture is thus almost insuperable. In this field, the boundary between valid scientific speculation and poetic fantasy is not well marked – not because anthropologists lack the 'scientific attitude' but because they are attempting to understand very deep and difficult phenomena.

Without precise and reliable theory, anthropology is in no position to predict the consequences of such political actions as propaganda for birth control or a reform of land tenure. Is it, then, merely a useless academic discipline, without practical benefit? The answer is, perhaps, that knowledge concerning the social customs of a particular group of men is of the greatest value in any relationship towards those men. Many of the crimes committed by 'civilized' nations against 'savages' have been due to ignorance, prejudice and misunderstanding concerning their native ways and cultural values. As an observer, it is not the duty of the anthropologist to intervene in political conflicts between different societies; but as an interpreter of one culture to another he can take some of the irrationality out of such conflicts. This is why 'Project Camelot', an abortive enterprise of the United States Department of Defence, was so misguided. It was, apparently, an attempt to make a detailed study of the factors that might lead to violent social change, insurgency and internal war in a country such as Chile. Sociologists, anthropologists, economists etc. were involved – not to do disinterested pure research, but to suggest how political action might best be taken against such change. This partisan project was thus a serious corruption of the neutrality of these academic disciplines and cast doubts on the credentials of hundreds of honest American social scientists at work in many foreign countries.

ECONOMICS

The only genuinely *quantitative* social science is *economics*. By attempting to reduce all social relationships to the single measure of money, economics aspires to the precision of physics. Since many of the actual relationships in an advanced industrial civilization are expressed as wages or prices, this is a reasonable scientific objective.

We may regard economics, in fact, as a science that has grown out of a technology. For many centuries, there have been merchants, bankers and royal treasurers who have had to practise the craft of

285

bookkeeping, have learnt the art of laying out their funds to the best advantage. Economics is to bookkeeping as physics is to engineering. The financial managers of a manufacturing business must, from experience, know how to deal with factors such as property values, rates of interest, tax reserves, which are determined ultimately by economic forces: the engineer designs his machinery by practical rules, which are subject, eventually, to the laws of physics. An economist is not necessarily a competent financier, just as a physicist is not necessarily a good engineer; but in each case the theoretical science would become sterile if it did not have some links with the worldly technique.

Economics is, therefore, a hard-edged science, with genuine predictive power in the field of social action. The prime example of this power, in our own era, is the economic theory of John Maynard KEYNES (1884–1946). The great economic depression of the 1930s was a disaster comparable to a great war for the people of almost all capitalist nations. Tens of millions of men were unemployed, and the families of the poor were in terrible distress. KEYNES argued, as a deduction from high economic theory, that this evil could be avoided by a deliberate change in government policy – in effect by reversing the principles of ordinary family finance, and spending heavily in bad times. This policy works: whatever may now be thought of the detailed theoretical justification of Keynesian economics, the serious problem of economic depression seems to have largely been solved by an application of this branch of social science.

Observe, however, that although KEYNES used mathematical reasoning he could not *calculate* the consequences of the policies that he proposed. Being himself not only an academic economist but also an exceedingly skilled practical financier, he had a sure grasp of the realities on which he based his theoretical arguments; but his mathematical equations contained many imponderable terms like 'liquidity preference' which could never have been measured directly. Economic forces depend, in the end, on human judgements, values and preferences: there is no absolute method for weighing a holiday in Majorca against a new washing machine. Genuine controversies amongst economists are not about details of algebra, but about the values to be assigned to various political and social ends, and cannot be resolved by the introduction of yet more terms in the equations. In the realms of general theory, the analogy with physics breaks down: unlike the fundamental laws of nature, economic laws are largely what men make them.

But in the search for scientific rigour in the social sciences, we might take notice of two new techniques of economic theory. The *theory of games* is an attempt to set up very simple models of some types of economic system and to study very carefully their exact behaviour. A model of economic competition, for example, might consist of two 'players' who make alternative 'moves' to gain 'points' according to some pre-arranged rules: the mathematical problem is to discover the optimum 'strategy' for each player, taking into account the most likely moves of his opponent at each stage. The study of such models, which was initiated by John VON NEUMANN (1903–57), one of the greatest applied mathematicians of our time, led to many new mathematical results, and in some cases to theoretical understanding of such phenomena as the formation of coalitions between groups of players, which seem to mirror those of real economic and political life. This is clearly a very instructive procedure for testing the inner logic and self-consistency of a conceptual model of a social system, but does not prove that events will actually turn out as the model seems to predict. The application of the theory of games to problems of military strategy (p. 331) is no doubt an interesting intellectual exercise which tells us a good deal about, for example, the central assumptions of the strategy of mutual deterrence, but cannot be taken seriously as a scientific guide through the terrible labyrinths of diplomacy and war.

The other new technique of economics is *Input–Output Analysis*. A 'matrix' is constructed to show the flow of goods and materials from one industry to another. Suppose, for example, that £50m worth of steel is used each year to make motor cars; at the intersection of the row labelled 'motor industry' with the column labelled 'steel industry' we write this figure. If sufficient statistical data can be collected to fill in this matrix for several hundred rows and columns, then we have a very coherent map of the whole economy. Manipulation of the numbers – in practice, with an electronic computer – provides numerical answers to many practical questions concerning the likely effects of new investment, changes of taxation, etc. It is difficult to say whether this is a brilliant new application of academic economics or a refined exercise in the ancient art of accountancy, but it certainly now plays an important part in political and economic planning. Unfortunately, the quantitative data that go into the equations are usually very imprecise, containing large sampling errors and many empirical scaling factors. The results of the calculations cannot be trusted very far in practice: even the crudest predictions may be quite wrong. We must always remember, moreover, that bookkeeping only describes the facts as given and tells us nothing about how men behave or might be expected to behave under different historical circumstances.

SOCIAL STATISTICS

Is it possible to obtain other genuine quantitative data concerning society? Even if we cannot *measure*, we can always *count*. The counting of people in a census, records of births, marriages and deaths, the collection of statistics of incomes, profits, rents, sizes of farms, numbers of bathrooms, school examination grades, robberies, suicides, hospital beds etc., are very large-scale activities of many public and private organizations. Much of this information is used for purely practical purposes, to check that a market exists, that work has been done, or as a measure of the efficiency of an existing service. But one can seldom distinguish between the technological, political and scientific uses of such data. Consider, for example, one of the earliest and most famous of 'social surveys' – Charles BOOTH's (1840–1916) *Labour and Life of the People, East London*, published in 1889. Using reports of School Board visitors, BOOTH was able to collect income and employment data on all the families in the area. His scientific purpose was to show the connection between conditions of employment and poverty, especially amongst the 7.5% of the population whose whole livelihood depended on casual earnings. But he demonstrated so effectively the extent of serious poverty, and the impossibility of alleviating conditions by private charity, that his work became an important political influence on legislation for old age pensions, health insurance, labour exchanges etc. The real value of such statistical studies is not in the confirmation of sociological theories but again in the elementary facts that they provide about the actual circumstances of people's lives. By emphasizing the need for reliable, objective data concerning individuals and institutions, the social and behavioural sciences exert a major influence towards realism and rationality in government and politics.

The difficult question is whether it is possible to draw useful theoretical conclusions from the analysis of such data. By applying statistical techniques and calculating the correlation coefficients between observed frequencies of occurrence of various individual characteristics can we make significant generalizations about the nature of man or of society? For example, some twenty years ago the correlation between heavy smoking and diseases of the lungs was observed to be very high: there seemed to be a much higher frequency of deaths from lung cancer amongst men who smoked several packets of cigarettes a day than amongst non-smokers. From this statistical observation it was inferred – almost certainly correctly – that smoking could be an important *cause* of lung disease. It is well known, and logically correct, that no causal relation of this kind can be *proved* by statistical evidence; yet most scientists would agree that this was a perfectly good piece of scientific argument. Can the same weight be given to similar observations of high correlations between social, behavioural or biological characteristics, in circumstances where it is impossible to conduct a direct experiment to prove the point?

There can be no mechanical answer to this sort of question. No amount of ingenuity in the manipulation of the data can really tell you whether the observed association is a direct causal link, or whether it is due to chance, or to faulty sampling or to an unobserved common factor. The essential point is to think about and to investigate the actual mechanism of the connection. The link between smoking and lung cancer is only too obvious – the inhalation of chemically active smoke particles into the lungs – yet some academic statisticians took an almost perverse delight in inventing all sorts of peculiar mechanisms that might possibly have produced the same statistical result. It was argued, for example, that people with an innate tendency to develop lung cancer might also have an unusual craving for nicotine – which is not much more unlikely than many accepted biochemical or medical phenomena. Strictly speaking the proper use of statistical methods is to *test* theoretical hypotheses, not to generate them out of the computer. The statistics clearly falsified the hypothesis that 'smoking cannot be a cause of lung cancer' but that could only be the beginning of a much more detailed investigation.

An example of the proper use of statistics in sociology is the central argument of the famous book on *Suicide* by Emile DURKHEIM (1858–1917). A natural hypothesis is that suicide occurs when a man is in a depressed state of mind, due to personal misfortunes such as bereavement or bankruptcy. One would expect, therefore, that the rate of suicide would be higher during wartime, or during an economic depression, than in more 'normal' times. The statistics showed otherwise. The incidence of suicide actually *falls* during a war, and can rise just as high in a period of boom as in a slump. This falsification of naive expectation was a very surprising sociological discovery and led DURKHEIM to a completely new way of looking at the relation of the individual to society. It must be admitted, however, that his further observation that the suicide rate is much higher amongst Protestants than amongst Catholics has been interpreted in many different ways without arriving at any certain conclusion.

SOCIAL INDICATORS

One of the difficulties even in counting 'objective' social phenomena is to find a proper way of defining categories and adding the numbers. Suppose, for example, that we want to construct a 'social indicator' for the amount of 'crime' in a given country or region – perhaps to show that a particular political policy has favourable effects. From the police records we draw data showing the number of 'indictable offences' for each year from 1920 to 1970. This number is seen to be rising more rapidly than the growth of population: does this mean that people are becoming less law-abiding? Not necessarily. It may mean that the police records are being compiled more accurately. Or it may be that the police are charging a higher proportion of criminals. Or perhaps the law has been changed, so that the category of 'indictable' offences now includes many that were previously regarded as minor infringements of the law. It could be that the rise has been due to the inclusion of very large numbers of motoring offences, which are by no means the same sort of social phenomenon as burglaries or drunken brawls. Murder is obviously a very serious crime; should it be given much greater weight in the total 'crime index' than petty theft? But a high proportion of murders occur in domestic circumstances, and have nothing to do with organized crime:

how should these be counted in the total? The labels that we give to particular types of social event are not themselves objective, and depend very much on history, prejudice, local custom and practical procedures. No amount of statistical analysis of miscellaneous data can bring them back to life and draw from them the social realities that they are supposed to represent.

It is an amusing – indeed devastating – commentary on the attempt to construct 'social indicators' that nobody seems to agree on the same formula. Some sceptical American sociologists, Bonjean, Hill and McLemore, in 1967 counted 3609 attempts, in the sociological literature, to measure various phenomena by the use of scales or indices. Of the 2080 different measures used, only 589 were used more than once, and only 47 were used more than 5 times. Such indices do not, therefore, satisfy the criterion of 'consensuality' which is fundamental to the scientific approach.

OPINION POLLS

These objections apply particularly to the study of 'attitudes'. The technique of the pollster with his questionnaire is all too familiar. 'What do you think of the policies of the present government?' (Strongly approve: approve: don't know: disapprove: strongly disapprove.) Skilfully conducted, such a poll can predict with reasonable accuracy the outcome of an election in a few days' time, but it is a nice question whether the results can be elevated to the status of scientific data. The point is not whether the resulting percentages are a reliable measure of opinion concerning the issues being studied, but whether such opinions can be made firm elements in valid generalizations about society. Do we learn more from them than can be deduced by a skilful and experienced reporter?

Consider, for example, the following data taken from *The American Soldier* (chapter XII, p. 380, vol. II), an account of the many surveys of attitude, opinion and experience made by the Research Branch of the US Army, during the Second World War. This table (Fig. 12.1) shows the relationship between the amount of combat experience of each bomber crew man and his answer to the question 'In general, what sort of physical condition would you say you are in at the present time?'

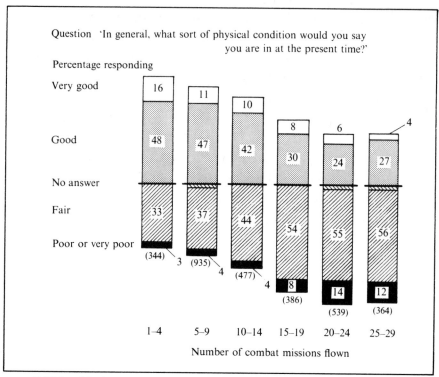

Question 'In general, what sort of physical condition would you say you are in at the present time?'

Percentage responding

Very good
Good
No answer
Fair
Poor or very poor

| | 1–4 | 5–9 | 10–14 | 15–19 | 20–24 | 25–29 |

16 / 48 / 33 / (344) / 3
11 / 47 / 37 / (935) / 4
10 / 42 / 44 / (477) / 4
8 / 30 / 54 / 8 / (386)
6 / 24 / 55 / 14 / (539)
4 / 27 / 56 / 12 / (364)

Number of combat missions flown

12.1 Relationship between amount of combat experience and answer to the question 'In general, what sort of physical condition would you say you are in at the present time?'

Evidently, the proportion of heavy-bomber crew members who regard themselves as in good physical condition falls from 64% at the beginning to about 31% at the end of a 'tour' of thirty combat missions. This is scarcely surprising: but notice also that the proportion stays steady, or even rises slightly, during the last third of the tour. This can be interpreted: men that have survived so long look forward now to the possibility of safely ending the tour and desperately hope to get it over and done with as soon as possible. Here is an authentic example of the complex interaction between social circumstances and personal psychology, worthy of serious scientific attention. Yet anyone who has read Joseph Heller's brilliant, satirical and sardonic novel, *Catch 22*, will recognize this as just one amongst many subtle patterns of thought, rational and irrational, resorted to by men in similar circumstances. The style of the novelist is not, on the face of it, scientific; but what he has to report is just as significant, for the student of human institutions, as the statistical results of questionnaires devised by earnest social psychologists.

Another example drawn from *The American Soldier* illustrates the *use* of attitude surveys at their best. At the end of the War, the problem was to devise a policy for the demobilization of the enormous number of men in the US Forces. Should priority go to men who had been in the army longest; or to men who had served overseas; or to older men; or to those with dependants? The wishes of the soldiers themselves were consulted by sample questioning, and a system of 'points' was devised giving appropriate weight to each of these factors – 1 point per month in the army, 1 point per month overseas, 12 points per child and 5 points per campaign star or combat decoration. The men with the highest points were then demobilized first. This policy proved remarkably successful. The men thought it was fair, and the whole operation was carried through without riots or unrest. In this sort of way, as in market surveying, political polling etc., social science provides a valuable new *technology*, by which the actions of large organizations can be guided into tolerable conformity with the wishes of those affected. Some people believe that this same science has given really effective tools to those who plot to *manipulate* public opinion for their own sinister purposes. But the evidence for this is weak: the demagogue, the political dictator and the huckster are not modern inventions, and do not need to know theoretical social psychology to make their way in the world.

PSYCHOLOGICAL TESTING

Social systems are inevitably complex and unpredictable: can we find better material for a powerful scientific discipline in man as an individual? A particularly interesting question is the nature of *intelligence*; the faculty that distinguishes the human being from other animals. We cannot *define* intelligence: but perhaps we can recognize and *measure* it as a basis for a quantitative science.

The notion that intelligence should be considered an innate characteristic of a particular man is little more than 100 years old. In his famous book *Hereditary Genius* (published in 1869) Francis GALTON (1822–1911) showed by a number of examples that high intelligence seemed to run in families, and must be a hereditary trait. He argued, along Darwinian lines, that it must be subject to genetic variation and to the mechanism of natural selection. In fact he jumped to the conclusion that this 'quantity', which he had no means of actually measuring, must be 'normally distributed' – i.e. that there must be about the same small proportion of 'geniuses', far above the average of intelligence, as there were 'idiots' the same distance below this hypothetical level.

Around 1900 the attempt to measure intelligence began in earnest. The pioneers were Alfred BINET (1887–1911) in France, and Charles SPEARMAN (1863–1955) in Great Britain. BINET was faced with a practical problem – how to pick out mentally defective children who ought to receive special schooling, and to distinguish them from those who had been labelled 'backward' as a result of faulty education. He devised a series of simple standard mental tests which could be administered individually by psychologists and teachers, and found that these tests could be graded fairly reliably according to the age at which they could be passed by normal children. He could thus say of a child of ten who was just able to pass the test for normal eight year olds that he had a *mental age* of eight, and was able to predict with reasonable certainty that this child would, within two years or so, be able to pass the ten year old test. He thus established the existence of a regular sequence of mental development that was followed step by step at a rate that varied from one child to another. This was, indeed, a significant scientific discovery.

SPEARMAN's work was more theoretical. The problem was to explain the fact that individuals get different scores in different sorts of mental tests: some people are good at arithmetical tests and bad at verbal tests, and so on. By elaborate statistical analysis he showed that these scores could be interpreted as the resultant of various mental 'factors' – arithmetical, verbal etc. – together with a major component that could be described as 'general' intelligence. Spearman's conclusions were probably somewhat overstated and much of the subsequent research using *factor analysis* is grotesquely inflated mathematically. A natural scientist would question whether genuine invariant factors (i.e. psychological 'forces') could be separated out by this somewhat arbitrary statistical technique applied to such very empirical data. But the various correlations between test scores are interesting qualitatively and have been fruitful in suggesting hypothetical schemes for the connections between various types of mental ability.

Whatever its theoretical merits, intelligence testing showed its value in practice during the First World War. The US Army *Alpha Test*, although a rather coarse group testing, was quite efficient at detecting and rejecting recruits of very low intelligence. This was the forerunner of innumerable batteries of intelligence and aptitude tests, now administered at the rate of several hundred million tests a year to aspirants for educational advancement, for entry into professional training, for selection as airpilots, for employment in the post office, etc. The predictive power of most of these tests has not been determined with great accuracy, but so long as both testers and those tested believe in them they have the advantage of appearing objective and unbiased.

EDUCATIONAL SELECTION

A familiar British example of the use of intelligence tests is in selection for secondary education. In the 1920s the conventional examination techniques for choosing children for scholarships for grammar schools came under attack. These examinations were shown by psychologists to be very haphazard and unreliable in their results. New standardized tests in English and arithmetic were devised, together with a general intelligence test which was found to be the best single predictor of grammar school success. Enlightened education authorities began to rely mainly on these tests which were felt to be the fairest way of choosing the most intelligent and promising working class children for more advanced education.

The 1944 Education Act provided facilities for secondary education for all children, and greatly increased the numbers going into the more academic grammar schools. The pressures of popular demand forced the general adoption of a more or less standardized procedure for selection at '11-plus' using intelligence and performance tests. Although the administration of such tests on a large scale became rather mechanical and although there were some serious technical defects in this procedure, such as the neglect of allowances for the range of age of one year around '11+', the whole system was regarded as a conscientious attempt to discover the children who were 'likely to benefit' by an academic education. Intelligence testing for educational selection was strongly defended by teachers and educational psychologists as a progressive and scientifically well-founded instrument of social justice.

Yet by about 1960 the whole policy was falling into disrepute, not only because many people did not like their children to be excluded from the better schools but also because a fundamental fallacy was becoming apparent. BINET's concept of mental age had been transformed arithmetically into the notion of an *Intelligence Quotient* (IQ), which is simply the ratio of apparent mental age to actual chronological age. This single number – 100, by definition, for the average child – was taken as a measure of innate intelligence: it was assumed that an IQ measured at the age of 11 could be treated as a reliable index of *future* intellectual ability as a student or adult. To a considerable extent, this assumption is valid; but careful experiments had shown that 20% of children gain or lose between 10 and 14 IQ points between the ages of 11 and 18. This variation is large enough to be significant in educational policy. It shows clearly that rigid rules for transfer into quite separate styles of education (with corresponding consequences for future careers, future earnings, future social class) are unjustified. As a result of this rethinking, educational psychologists now favour a policy of 'comprehensive' secondary education, where each child would have the opportunity to develop more or less academically as it passed through a single school.

This episode is instructive, because it shows the danger of relying too heavily on 'science' as a guide to action. Nobody doubts that intelligence tests provide a rough and ready guide to relative competence in solving the sorts of problems that are being tested, at the time of testing. But there is little justification for treating IQ as a 'quantity', comparable with a physical property such as density or temperature. What could this ratio mean for an adult, whose chronological age continues to increase after his 'mental age' has come to a halt? Following, presumably, old GALTON's conjectures, psychologists adopted a convention of cooking the scores of intelligence tests so that they appeared to be 'normally' distributed when measured on a large population – in fact, this entirely arbitrary criterion is often used as a definition of a 'good' test.

Another type of criticism is that an IQ score may depend as much on cultural experience as on some innate 'biological' capacity. Middle class children, with a more varied cultural background and the opportunity to learn a larger vocabulary, are inevitably at an advantage. The whole purpose of intelligence testing was to eliminate any bias due to education or class, but this seems an unattainable goal: can one imagine such a thing as 'intelligence', in an adult human being, that was not strongly determined by the whole process of his upbringing in some specific society. The typical puzzle questions of even the most cunning of tests are governed by the very notion of a 'puzzle', as seen by an adult who has already shown his own competence in that very special way of life that we call a university. Would an intelligent Eskimo see such puzzles as stereotypes of the 'problems' that *he* has to solve in *his* cultural context.

RACE AND INTELLIGENCE

This leads us into the ancient problem of 'nature' or 'nurture'. To what extent is intelligence a genetic trait? Are the children of intelligent parents themselves intelligent through their bodily inheritance, or is it because they have been brought up in a more stimulating family environment. From studies of identical twins that were separated at birth it has been argued that something like 70–80% of the variation of intelligence from one adult individual to another may be accounted for genetically, and that differences of environment can only have an effect of a few per cent. But this is well established only for differences of up-bringing and education within the same general pattern. More sophisticated statistical studies suggest that the true genetic factor may be only about 50% of the total variance of IQ. Nobody knows – and it is very difficult to think how to prove – whether differences of environment as large as those between different ethnic groups in a polyglot society can be the sole cause of observable differences in the average test scores of their members.

And now, of course, we find ourselves in the middle of one of the most controversial and sensitive questions in the whole sphere of social relations: are there significant differences of 'innate intelligence' between different 'races' of mankind? The difficulty of giving an objective quantitative measure to the concept of intelligence is apparent: is the concept of 'race' any more precise?

If we go back again to mid-Victorian times, we find a genuine scientific interest in the biological differences between human groups. The discipline of 'physical anthropology' was devoted not only to the characterization of various groups according to the colour of their skin but also to the differentiation of the very mixed-up inhabitants of Europe according to such refined indices as the shape of the skull. The classification schemes that were worked out by various enthusiastic research workers, like those for many genera of plants and animals, did not always agree in detail, but at least the questions that were being asked were capable of falsification by direct observation. Much of this work may now seem too speculative, but the modern study of the relative frequencies of blood groups in different racial stocks can be tied closely to molecular biology, genetics and precise quantitative concepts such as genetic drift.

The trouble was that the social and cultural characteristics of each racial group were also regarded, without proof, as outward manifestations of innate, biologically heritable traits. In particular, the cultural dominance of the white races was attributed to genetic superiority of intellect, character etc. This view was obviously mixed up, in many cases, with the attempt to justify political domination, as with American slavery and European colonialism. But in the mind of a man like (once more!) Francis GALTON, it would be difficult to separate this sort of prejudice, born of his personal experience in South Africa, from the theoretical rationalizations of 'social Darwinism', whereby it was argued that the white races must be more 'fit' to rule because they had survived hardship and evolved into a superior form.

This is an important point because it illustrates an alarming and dangerous consequence of scientific progress. A new and perfectly good scientific theory, such as the principle of biological evolution by natural selection, is taken out of its proper sphere – the origin of zoological and botanical species – and applied uncritically in support of all sorts of dubious notions. The idea of a struggle for existence and of survival amongst competing cultures or systems of social behaviour is not utter nonsense; but as soon as one looks at it critically one sees so many difficulties and objections that it is impossible to accept it as a necessary truth.

The modern controversy concerning race and intelligence is an intellectual relic of this old topic. The attempt is made to *prove*, by statistical analysis of batteries of intelligence tests, that the old conjectured differences really exist. Again, it is not a completely absurd hypothesis, but the more one considers the technical problems of demonstrating the effect beyond all doubt, the more difficult they seem. With 'intelligence' itself very far from being a culturally neutral concept, what would one be saying if one simply pointed out that blacks, on the average, score ten points less than whites on a given set of puzzles? In any case, the range of variation amongst individuals in each group is so large that one could never make this piece of 'scientific knowledge' the basis for an educational policy.

So formidable are these obstacles to the 'scientific' study of the relations between race and intelligence that one must begin to question the motives of those who advocate such a programme. Are they truly honest seekers after truth or do they have political, ideological ends which they are attempting to justify by an appeal to scientific authority? It must not be forgotten, that the whole investigation is offensive, by implication, to those who might thus be deemed inferior; why should their sensibilities be ridden over in the name of research when the most likely outcome is continued doubt.

IS SCIENCE NEUTRAL?

Some people go much further and use this controversial topic as evidence that science can never be 'value free' and 'neutral'. They point to the power of the social ideologies within which we are all brought up, and assert that every scientist must be affected by prejudices of capitalism, or socialism, or fascism, or whatever it is, in all his observations and deductions. Such an ideological slant is particularly significant in the social sciences, where there is no simple instrumental device to correct the bias of the observer. Looking back to such famous works as those of MALTHUS (p. 130) or SPENCER (p. 283), we can now easily detect implict assumptions about the organization of society that would have seemed self-evident at the time but which we now reject. It is impossible, therefore, for any one of us now to avoid making the same sort of error, which will thus spoil every attempt at scientific 'objectivity'.

This fundamental challenge to all efforts to build up a valid body of social science has much force. It is certainly to be taken very seriously whenever we try to use 'scientific' knowledge as a basis for political decisions that touch upon human values. But this is only an extreme and significant form of the basic question of the absolute 'truth' of scientific knowledge of all kinds. In the end we are bound to rely upon our own judgement of the validity of each observation and deduction, and are all subject to the intellectual climate of our own times. The paradigm of classical physics in the early twentieth century was as much an obstacle to the acceptance of relativity and quantum theory as the cultural paradigm of European society was to the acceptance of anthropological relativism and functionalism.

Having seen this, however, the gravest folly is to suppose that the bias of an objectionable (i.e. false) ideology can be corrected by embracing its opposite. In many many cases, those who denounce science for not being neutral are mainly concerned to make it an ally on their own side of some political or religious conflict. But this is a matter on which each one of us must exercise his own judgement, relying upon the common sense virtues of honesty and scepticism that he has derived from his personal experience of life.

THE SOCIOLOGICAL POINT OF VIEW

This chapter was begun with little hope of settling so vast an issue as the place of the social and behavioural sciences in society. It might be summed up somewhat as follows:

The sociological viewpoint, that takes man in society as a proper object of unprejudiced enquiry, is immensely valuable. It removes the blinkers of conventional religious, political and social systems of thought, and gives the observer freedom to see things in a new light. The map of the social and psychological world that the average person carries around inside his head is extraordinarily narrow and restricted: sociology and anthropology suggest entirely different schemes by which it may be represented.

The uncovering of the immense variety of actual social and cultural conditions is also extremely instructive. It is very easy to be blind to the realities of life around us, to ignore what is not before our eyes, and to notice only those people and events that our prejudices allow us to see. By observing, recording and counting all manner of simple facts, the social scientist brings them with due weight to the attention of the managers and politicians. Scientific criteria of objectivity, or reproducibility of data, of measurement, and of quantity, make this information much more reliable, as a basis for possible action, than mere opinion.

But the part played by theory in this realm of science is much weaker than in the physical and biological sciences. The facts themselves are often sufficient to demolish what J. R. Ravetz has called 'Folk Science' – the old systems of prejudice and superstition that govern the minds of most people concerning race relations, sexual practices, barriers of caste and class, etc. But the social sciences have not succeeded in building up a thorough-going theoretical framework from which reliable predictions can be deduced in the abstract without reference to the facts. The real power of physics is that we can fire a rocket into space, with prescribed velocity, and can calculate with precision that it will land on the moon. Nobody can launch a social reform or an industrial enterprise, and calculate *its* trajectory by the aid of mathematical sociology or economics. Theoretical models and concepts are absolutely essential aids to practical research in the social sciences, but they can seldom be taken literally as guides to action. Little is gained by setting aside a Folk Science shrewdly based upon long experience and replacing it with a pseudoscientific 'expertise' based upon glib theorizing and superficial observation.

On many matters of great social concern, true science has nothing positive to say. We do not know, for example, whether black men are somewhat cleverer than white men or the other way round. What society needs is the positive assertion of our *ignorance* on such issues – that is, the denial of any scientific authority that is claimed to favour a particular prejudice or superstition. Perhaps this is the most important ideological task of the scientific attitude in the realm of general ideas, social action, philosophy etc. – to make clear the values of scepticism, open-mindedness, and attention to ascertainable facts, on matters that cannot be settled by the mechanical application of the type of rationality used in science itself.

13 SCIENCE AND WAR

There were all sorts of things that people were routing out and furbishing up; infernal things, silly things; things that had never been tried; big engines, terrible explosives, great guns. You know the silly way of those ingenious sort of men who make these things; they turn 'em out as beavers build dams, and with no more sense of the rivers they're going to divert and the lands they're going to flood!   *H. G. Wells*

We talk often of science responding to the *needs* of society. The most persistent *demand* is for military power. Whatever we may think about it from an ethical point of view, it cannot be denied that preparations for war, and war itself, are an integral part of modern human life.

The sociology of war between nation states is seldom discussed, except from the standpoint of the political historian. The role of science and technology in modern war is all too evident, and much of the abhorrence of war amongst civilized people has been transferred to science itself, without deeper study of the underlying causes. This is too large a theme to be discussed here in full. We shall look at this evil and horrifying phenomenon only as it affects 'pure' science, and the scientific community.

Technical mastery has, of course, always been a decisive factor in warfare. But usually this has been a practical art, improved and developed by trial and error. Some of our best examples of sophisticated technology (see, e.g. p. 10) have been, quite simply, weapons of war – siege engines, swords, cannon etc. More general techniques, such as shipbuilding, architecture, aircraft design and mechanical engineering, have been strongly influenced by, and immensely influential in war. The East Indiaman and the ship of the line; the cathedral and the castle; the jet bomber and the jet airliner; the caterpillar tractor and the tank; they are closer than cousins for the man who designs and constructs them. No serious history of technology, from the very earliest times, can ignore the military applications of peaceful inventions, nor the influence of prime military necessity as the mother of many useful inventions. Medicine itself, that most humane of all technologies, owes almost as much to war as it does to peace (pp. 147–65). But this is not our present theme.

We would also discuss war as a source of inspiration for pure science. It is no accident that the science of mechanics, with its peculiar emphasis on the motion of spherical projectiles, grew up during the Renaissance, the great era of the cannon. GALILEO, for example, was deeply influenced by what he could observe in the great arsenal of Genoa and RUMFORD's research on heat (p. 75) stemmed from his post as Minister of War to the Elector of Bavaria. Scientists have never been reluctant to draw research material from the experience of war, nor to add their own personal contributions to military practice. Some 10% of the research undertaken by the Fellows of the Royal Society in its earliest years was related, directly or indirectly, to military technology (p. 19).

But the wholesale enrolment of the scientific community on behalf of the nation at war is a modern custom. Of course, there was ARCHIMEDES (p. 10), whose brilliant devices frustrated the Romans in their siege of Syracuse: but he just happened to be that sort of genius, and felt bound to help his patron against his enemies. As Plutarch records

Archimedes possessed so lofty a spirit, so profound a soul, and such a wealth of scientific inquiry, that although he had acquired through his inventions a name and reputation for divine rather than human intelligence, he would not deign to leave behind a single writing on such subjects. Regarding the business of mechanics and every utilitarian art as ignoble and vulgar, he gave his zealous devotion only to those subjects whose elegance and subtlety are untrammelled by the necessities of life.

This high-minded disdain for the lowly art of killing other men has remained the official doctrine of pure science until our own day, being matched only by the scorn of the professional soldier for the crackpot inventions of the academics.

The change came in the First World War. On both sides, the academic scientists were eventually mobilized. For example, the leading British and French professors of physics, W. H. BRAGG and P. LANGEVIN, formed a committee to work on antisubmarine detection: LANGEVIN found that piezo-electric microphones were efficient at detecting submarine engines. Again F. A. LINDEMANN (see p. 140), G. P. THOMSON,

13.1 H. G. J. Moseley in uniform.

E. D. ADRIAN, and G. I. TAYLOR could all be found together at the Royal Aircraft Establishment at Farnborough, designing and flying new aircraft. The mobilization of scientific talent was slow to get under way – the loss of MOSELEY, one of RUTHERFORD's most brilliant pupils (Fig. 13.1) who was killed in action at Gallipoli, like millions of other young men, was the convincing argument for official action – but it was deliberate and it paid dividends.

CHEMICAL WARFARE

The most spectacular and successful new weapon produced by the scientists was poison gas. The idea was not new; as far back as 1812 Admiral Lord Dundonald had suggested the use of burning sulphur as a weapon. It was again proposed in 1855 during the siege of Sebastopol, but was considered too horrible. Chemical warfare was actually outlawed by the Hague Convention of 1899, to which all the major powers sub-scribed.

But in 1915 there was near deadlock in trench warfare on the Western Front. The French had experimented half-heartedly with tear gas grenades in 1914, and the Germans had tried shrapnel containing a chemical irritant. In early 1915 they tried another chemical agent against the Russians, but the weather was too cold. The technical 'break-through' came on 22 April 1915, near Ypres. Two hours before sunset on that fine spring day, high explosive fire against the French trenches was halted, and 500 cylinders containing 168 tons of chlorine gas were opened. The effect was devastating (Fig. 13.2). The cloud of gas produced 15,000 casualties, and all resistance was eliminated for several miles deep over a four-mile front. If the Germans had been prepared to attack on a really large scale, they could have broken right through the trench lines. Within a few days, however, simple gas masks became available (Fig. 13.3), and the new weapon, which might have won the war, was contained.

Who was responsible? The leading German chemist of the day was Fritz HABER (1868–1934), who had invented the nitrogen-fixation process that made Germany self-sufficient in nitrates – a key material in the munitions industry. He became a captain in the army, along with

13.2   A gas attack in the trenches.

13.3   Some of the first gas masks. 2nd Battalion Argyll & Sutherland Highlanders, 19th Brigade, 6th & 27th Division, Bois Grenier Sector, March–June 1915.

W. NERNST (1864–1941), the great physical chemist who had been LINDEMANN's teacher. Together they worked on poison gases, developing the scientific techniques, taking command of the gas warfare organization and even watching the effects on the battlefield. For this patriotic effort HABER was later strongly criticized by the scientific community, which still had strong international sentiments, especially when he was awarded the Nobel Prize for his nitrogen-fixation process. Ironically, HABER was Jewish: in 1934 he lost his post, and died in exile in Switzerland.

But, of course, the British and French did not hesitate to reply in kind, and were soon making gas attacks of their own. New chemicals, such as phosgene, which is much more toxic than chlorine, were tried out, and new gas masks were designed (Fig. 13.4). Techniques for dispersion of gas by shells and mortar bombs were invented. A big research and development effort was begun, involving many good young scientists. For example, the war service of C. A. LOVATT EVANS (1884–1968), a distinguished physiologist, is recorded in his Royal Society obituary. After qualifying medically in 1916, he joined the Royal Army Medical Corps, and worked in the anti-gas department of the Royal Army Medical College at Millbank. There they studied arsine, phosgene and hydrocyanic acid. They also advised the use of mustard gas (which they named) but this was rejected by the army: when the Germans began using this chemical fifteen months later, the RAMC boys were very angry at being forestalled by the enemy. In fact, mustard gas is still regarded as a very effective and dangerous weapon because it acts on and through the skin, so that a mere gas mask is inadequate protection.

The scale of chemical warfare in the First World War is seldom realized. More than 3000 substances were screened as possible toxic agents, though only a dozen were found to be effective: 125,000 tons of gas were dispersed by 17,000 specially trained chemical troops: more than 9 million shells were filled with mustard, and produced 400,000 casualties – which was five times as effective as high explosives or shrapnel. It is estimated that gas caused something like one million casualties in the War: in the last year of the War, some 16% of British and 33% of American casualties came from chemical weapons. It was thus a significant military factor, which both sides used to the fullest extent, despite the popular attitude that it was 'dastardly', 'inhuman' etc.

13.4 Machine gunners wearing gas helmets. Battle of the Somme, near Orvilliers, July 1916.

But note that in the 1920s the general technical opinion was that gas was a 'humane' weapon. It was pointed out that the death rate per casualty was much lower, and that the permanent disability effects were less serious, than for conventional weapons. Thus, although chemical warfare was again outlawed by the Geneva Protocol of 1925, the scientific community could argue that it was not so wicked after all. During the First World War, it was the scientists who got on with the job of developing new chemical weapons and providing defence against them. Even J. B. S. HALDANE (p. 86), that champion of goodness and kindness, wrote a little book emphasizing the humane advantages of gas. It was the government and the politicians who took the emotional stand and engineered the Geneva protocol.

Despite its immense importance, even in recent years, I don't propose to trace the further history of chemical warfare. The situation was quite altered by the German invention of exceedingly toxic nerve gases in the 1930s, but somehow neither side dared to use them in the Second World War. Although the major powers stockpiled large quantities of gas, and prepared anti-gas defences on a large scale, the arguments of mutual deterrence prevailed. But all this now belongs to the history and technology of war itself.

What we learn from this episode is that in the stresses of an all-out 'patriotic' war there are no real ethical restraints on scientists or on any other citizens. Modern war uses the whole industrial machine, with all its techniques. The Germans made the first big gas attack, not because they were specially evil men, but because their chemical industry was especially powerful. The escalation of the scientific effort on both sides was an inevitable consequence, and in no time the full strength of the scientific community was being exerted. The boundary between the pure scientist and the technologist was shown to be meaningless: the 'academics' had reserves of knowledge and inventiveness that were scarcely tapped in peacetime and which could be decisive in war. The question of the relative 'humanity' of a weapon of a particular type was scarcely considered: military effectiveness was evidently the only criterion for its adoption. Notice also that, apart from HABER, the scientific effort was almost anonymous. The general public made no heroes out of its scientists in the First World War. As a Cabinet Minister was reported to have remarked: 'What I like about scientists is that they are a team; you don't even have to know their names.'

RADAR

Skip forward some twenty years, to the Second World War. This great conflict was dominated and decided by scientific techniques. Methods of warfare that could not have been thought of except within the context of pure scientific research became of overriding importance. No nation without a scientific establishment could have sustained its military potential.

Again, the whole subject is too large: let us look briefly at the story of radar.

The basic idea of detecting radio waves that had been reflected back to a receiver goes back to about 1924, when APPLETON and BARNETT (UK) and BREIT, TUVE and TAYLOR (USA) began to study reflections from the ionosphere. Occasional observations of the effects of aircraft or ships interfering with radio propagation were also reported and several people proposed that this effect might be useful for detecting ships in darkness or fog. In 1930 the US Navy began a project for the detection of ships or aircraft by radio, although it was not until 1934 that the idea of transmitting radio *pulses* was seen to be the key to the technique (Fig. 13.5).

13.5   Diagram illustrating idea of pulsed radar. (*a*) Problem: for aircraft warning on ships by radio, transmitter and receiver must be close together. (*b*) Idea: use short pulses with long spaces between, so transmitter will be 'off the air' while the receiver is listening for echoes.

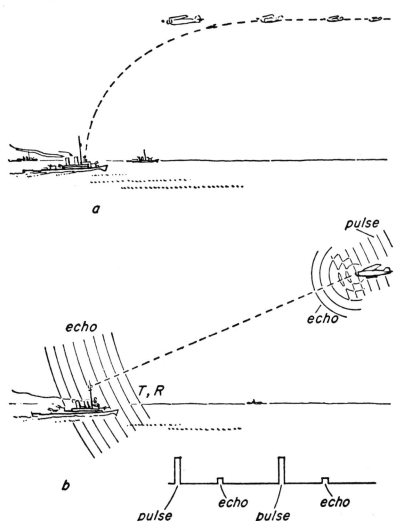

Much the same thought processes occurred in Germany, where similar developments were put in train. Once again, the basic principles were fairly obvious to any well-trained physicist or electrical engineer, and did not demand exceptional scientific or inventive genius.

Actually, with the radio techniques of the mid thirties, the production of an effective pulsed radar system was not a trivial problem. It was all too obvious that shorter wavelengths would be preferable, but these could not be generated efficiently at that time. Nevertheless, the appropriate circuitry for primitive devices working at a wavelength of a few metres (Fig. 13.6) was designed and built in about two years; by 1936 excellent echoes from aircraft at distances of 50–100 km were being obtained (independently!) in Britain, America and Germany. The German equipment was still at the experimental stage when she went to war but was later developed to a high technical pitch (Fig. 13.7). By the time the US entered the War in 1941 about twenty warships had fully installed radar systems (Fig. 13.8). Indeed, the Japanese aircraft attacking Pearl

13.6 Original 28 Mc Radar transmitter with synchronizing keyer. 17,000 volts on exposed wires and condensers on top shelf. Installed in fieldhouse. April 1936.

13.7 German radar equipment. The 'Giant Würzburg'. This photograph was taken by a member of the Belgian Intelligence Organization during the Second World War.

13.8 Radar antenna on US warship, 1938.

Harbor were detected in good time to sound the alarm, but the warning was disregarded.

The story of British radar is especially instructive. It seems that, in 1934, A. P. ROWE, an Air Ministry scientist, was leafing through the 53 files on air defence and was appalled at the little effort that had been made to tackle the problem scientifically. He suggested to his boss that a proper Committee on Air Defence should be set up to initiate and organize research in this field. This committee was chaired by Henry TIZARD, and got into a certain amount of political and technical trouble in its relations with F. A. LINDEMANN (p. 140). The essential historical event occurred before the first meeting, when H. E. WIMPERIS, Director of Research for the Air Ministry, asked R. A. WATSON-WATT (1892–1973), Director of the Radio Research Station at Slough, whether enough radio energy could be directed at an enemy aircraft to destroy it. Watson-Watt passed this query on to one of his staff, A. F. WILKINS, who did a 'back-of-envelope' calculation showing that the idea of a 'death ray' was nonsense, but that detection of the reflected radiation was theoretically feasible. A memorandum to this effect was passed back to the Tizard Committee, who encouraged WATSON-WATT, ROWE and a small team to begin secret experiments. By April 1937 the Government had decided to cover the east and south coasts of Britain with a radar network, which was completed in the summer of 1939 (Fig. 13.9).

This system cost about £1m – not a negligible sum in the Defence Budget of that era – but it was, in fact, technically very primitive. Because it worked at quite a long wavelength – about 10 metres – the shape of the beam was very irregular and each station had to be very carefully calibrated. There were big gaps, especially at low angles, which had to be plugged with further special equipment (Fig. 13.10). But the choice of a 'third best' system was deliberate. As WATSON-WATT remarked: 'The best never comes, the second best comes too late.' Perhaps also the improvised character of much of this equipment owed something to the scientists who built it; they were experienced in the use of experimental apparatus rather than in engineering design for industrial production.

On the other hand, the chain stations were carefully sited (Fig. 13.11) and connected by land lines to filter centres and thence to operations rooms, where plotters put on to a map the information received by

13.9   Chain Home radar mast.

13.10   Polar diagram of typical Chain Home radar station for reliable pick-up of single fighter aircraft along the line of sight.

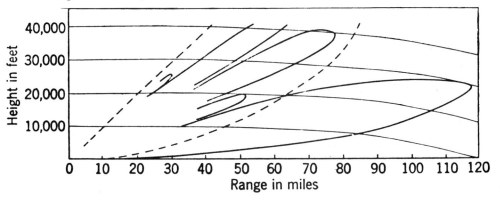

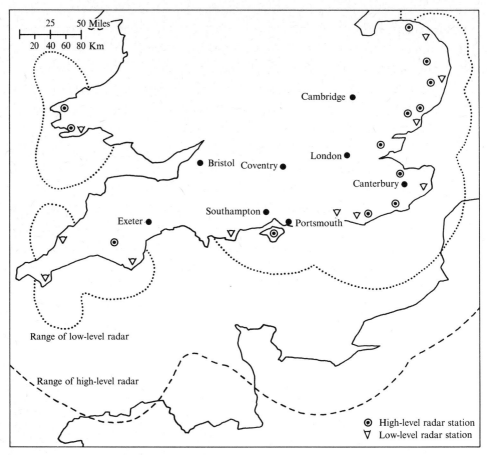

13.11 Map of Chain Home radar system.

telephone (Fig. 13.12). Instead of elaborate radio devices, computers etc. to correlate and correct the messages coming from the various stations, a 'filter officer' stood beside the plotters and decided, for example, whether two neighbouring tracks were identical. Thus human judgement was used to the full to get the maximum performance from the crude technical equipment.

Another important factor was that TIZARD insisted that the Royal Air Force should have thorough exercises in 1938 in the use of radar for air interception. When the war broke out, the operational officers were familiar with the technique, and already 'radar-minded'. Note also that Sir Hugh DOWDING, the Head of Fighter Command during the Battle of Britain, had been Air Member for Research in 1935 when the TIZARD committee was created, so he was already interested in radar and ready to

314

make it the basis for military operations under his command. These human links between the scientists and the fighting men were of great significance. Without this mutual respect and confidence, and without radar, the Battle of Britain could not have been won. When the Luftwaffe attacked, it was met at every point by British fighters, directed to the point of interception by the Chain Home system. The effective strength of the Fighter Command was practically doubled by not having to mount continuous patrols to detect the enemy. That was the measure of the power of the new weapon.

Yet the British lead in radar in that first decisive year of the War was not due to superior scientific skill in radio engineering, nor to the discovery of some unimaginable secret of nature. The significant factor lay in the wise judgement of a few high-ranking officers, politicians and scientists who were willing to work together without much regard for service protocol, who could respond both to unorthodox theoretical proposals and to the practical experience of the men of action, and who saw the absolute necessity of urgency and economy in the use of limited

13.12 WAAF plotters and filter officers. Operations room at HQ Fighter Command in 1940.

13.13   Randall's original cavity magnetron.

resources of men and materials. The scientists not only contributed technically: they also gave their rational realism, their freedom from conventional military doctrines and their ability to grasp the prime objectives of the whole operation.

The big technical breakthrough that maintained this initial lead came in the summer of 1940. As we have seen, the fundamental need of all radar systems was to go to much shorter wavelengths, so that a much more precise beam of radiation could be sent out and swept around the sky. This could only be achieved with some form of high-powered transmitter valve. The Germans believed that this was impossible; but in the autumn of 1939 several British teams, mostly consisting of academic nuclear physicists, began research on this problem. J. T. RANDALL, working at Birmingham University under M. H. L. OLIPHANT, tried the idea of building a resonant cavity into the interior of an existing microwave device, the magnetron (Fig. 13.13). The result was almost instant success – 10 kW of power at a wavelength of 10 cm – quite enough for a radically new radar system. Here was the beginning of the microwave technology which now dominates radar and telecommunica-

316

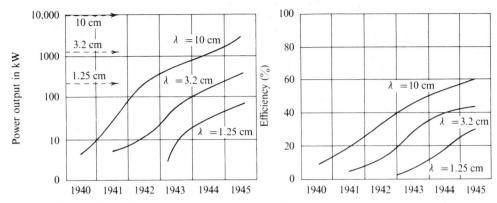

13.14   Historical development of microwave magnetrons.

tion engineering. The technical mission sent from Britain to the USA in September 1940 took with them a 'cavity magnetron'; this is said to have been the single most valuable product ever carried across the Atlantic. Here we see the sort of contribution that fundamental academic science can make to war – a revolutionary 'secret weapon' produced in the midst of the conflict, capable of altering the tactical balance of power in a wide variety of military operations.

But the exploitation of the breakthrough made further demands on the scientific community of Britain and the United States. Many new technical tricks had to be discovered before the microwaves could be handled efficiently (Fig. 13.14). Many different types of radar had to be developed for anti-aircraft gun laying (Fig. 13.15), night fighter inter-

13.15   Diagram of radar operating with a gun battery.

SCR-584 WITH GUN BATTERY

ception, U-boat detection, bombing, naval detection and gun-laying etc. A particularly difficult job was the design of a proximity fuse – a radar set that could be carried by an artillery shell and explode it when it was within a short distance of its target (Fig. 13.16). We take for granted the Plan Position Indicator, by which the rotating aerial of a radar set is synchronized with rotating sweep of the electron beam on the face of a cathode ray tube, thus tracing out a map from the radar echo (Fig. 13.17): this device is easy to think of, but required great skill to design and put into practical operation.

Nowadays this sort of development work would be done by electronic engineers, that is, technologists familiar with the basic methods and experienced in the practical design. But in 1940 these people did not exist: the completely new technology of microwave engineering was just

13.16   Radar-fused 155 mm shells bursting over target, France.

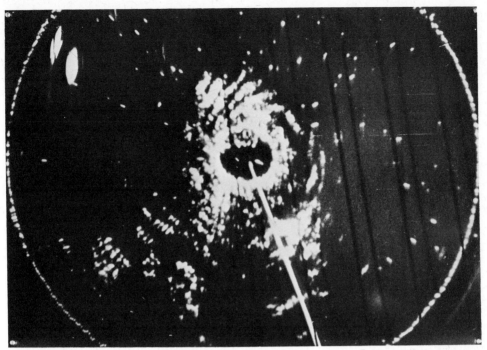

13.17 Early plan position indicator display of the city of Leicester, taken with H2S equipment, 1943/44 vintage.

in its infancy. Many academic research physicists from the universities were drawn into the new craft: in fact, a good proportion of British university physicists now in their 50s had this valuable experience during the War. This is one of the reasons why the application of microwave electronic techniques in all branches of the physical sciences became so fashionable after the War. The completely new science of radio astronomy was only one of the intellectual by-products of radar engineering.

The tremendous pressure to get things done created a special human atmosphere inside the radar R & D organization. In September 1939 various university scientists were taken round the radar research establishments and shown what was going on. As they were drawn into the work, they were not forced into harsh bureaucratic conformity, as they might have been in a regular military organization, but were treated as independent intelligent people. Many characteristics of 'academic' life were still preserved within the Telecommunications Research Establishment (TRE) that grew out of the little group of 1936–8. A peculiar feature

13.18 A 'Sunday Soviet' at Malvern College.

of the British radar research was the 'Sunday Soviet' (Fig. 13.18) where senior RAF officers and other 'important' people would take part in open forum discussions with quite junior scientists about operational problems and new technical devices. The scientific spirit of free debate amongst intellectual equals was thus maintained within the closed and secret development organization.

The practical achievements of the 'pure' scientists in the Second World War were thus of the highest order. Although they could at times have benefited from conventional engineering skills, they showed that they could do things far beyond the abilities and imaginations of the established military experts. Similar successes in the field of operational research, such as the work of P. M. S. BLACKETT and J. D. BERNAL against the U-boats, showed that scientific intellectuals were indispensable in modern war. For this reason they have continued to be employed by military organizations ever since. This is one of the basic facts about science in our time. No modern defence ministry can afford not to involve basic scientists on a large scale in its inner activities.

Remember, however, that the Second World War, on the Allied side, was absolutely patriotic. There was no question but that the idea of losing the War was quite unacceptable to everybody involved. The basic problem of morale – belief in the cause for which one was fighting – was thus solved, and the general (well-founded) belief that the Nazis would stick at nothing removed any inhibitions about the ethics of this use of science. Notice, for example, a remark by BERNAL in 1939 concerning a possible boycott of war research: 'In the present state of the world it is even doubtful whether such a policy would have good results, for the immediate first effect would be to put democratic countries at a disadvantage in regard to Fascist ones.'

It would again take us too far afield to discuss the contribution of German science to their war effort, although it must be said that this relied much more on immense engineering expertise than on academic physics and chemistry. The history of the Atom Bomb, with all its drama of research, development, military decision and espionage is also too complex (and too important in itself) to be summarized here in a few pages. Let us therefore move on twenty years to another episode illustrating the relationship of the scientific community to the making of war – the debate in the USA concerning the ABM – the Anti-Ballistic Missile System (Fig. 13.19).

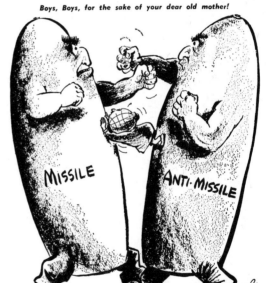

Boys, Boys, for the sake of your dear old mother!

13.19 Missiles and ABMs.

MISSILES AGAINST MISSILES

In 1955 the Americans were developing ground-to-air missiles for anti-aircraft defence and came to the conclusion that it was technically possible with such a weapon to hit an enemy ballistic missile as it approached its target area. The essential problem, which they believed could now be solved, was to construct radar sets of sufficient sensitivity

Table 13.1. *Number of fatalities in an all-out strategic exchange*[1]

	Soviets strike first, US retaliates		US strikes first, Soviets retaliate[3]	
US programmes	US fatalities (in millions)[2]	Soviet fatalities (in millions)[2]	US fatalities (in millions)[2]	Soviet fatalities (in millions)[2]
Approved	120	120+	100	70
Posture A	40	120+	30	70
Posture B	30	120+	20	70

[1] Assuming no Soviet reaction to US ABM deployment. Fatality figures shown here represent deaths from blast and fall-out; they do not include deaths resulting from firestorms, disease, and general disruption of everyday life.

[2] The data in this table are highly sensitive to small changes in the pattern of attack and small changes in force levels.

[3] Assumes US minimizes US fatalities by maximizing effectiveness of strike on Soviet offensive systems.

and accuracy to ensure interception. By 1961 the *Nike–Zeus* system, which would have provided batteries of short-range anti-ballistic missiles for the protection of the major city areas of the US, was at the prototype stage.

But it was obvious that this system would not really be very effective as a defence against Russian missiles, unless deployed on an extravagant scale, so it was never put into production. In 1963 work was begun on a more advanced ABM system, *Nike–X*: by 1965 it had apparently been confirmed that the enemy missiles could be intercepted and destroyed at very long range, outside the atmosphere, thus shielding a much larger area of the country and causing much less damage from fall-out. But the decision to go ahead with the installation of this system was still deferred.

Unfortunately, at about this time, reports came in that the Russians had devised an ABM system and were deploying it around Moscow and Leningrad. They said that ABMs were purely defensive anyway but the American strategy experts got very worried because the balance of 'mutual deterrence' might be upset if the Soviet Union need no longer fear terrible destruction from an American (counter) attack. This alarm was needless. The figures given in January 1967 by Robert McNamara, the US Secretary of Defence, showed that even a very expensive ABM system really made little difference: the estimates of reciprocal death and damage in all-out nuclear war were still too terrible to contemplate (Table 13.1).

Nevertheless, in September 1967, the US Joint Chiefs of Staff got their way. After an inconclusive Congressional debate, the political decision was taken to go ahead with a 'thin' ABM system, called SENTINEL, covering vast areas of the country. There was talk of providing defence against Chinese nuclear threats, but this scarcely added up to a compelling argument for embarking on such an expensive programme which would be certain to provoke further moves in the Arms Race.

At the heart of the SENTINEL system (Fig. 13.20) are several very

13.20 'SENTINEL' ABM system as originally planned.

13.21 Full scale mock-up of the SPARTAN missile.

13.22 SPRINT missile in flight.

13.23 Missile site radar for the SAFEGUARD ABM system.

324

large and sophisticated 'Perimeter Acquisition Radars' (PARs) whose job is to track the attacking missiles at long range. This device, costing hundreds of millions of dollars, can deal with many different objects at once by 'steering' its radar beams electronically from a vast array of aerials. In fact, a PAR set works rather like a radio telescope in reverse, and presumably benefited by research in that peaceful art. But one must remember that the attacking missiles would be accompanied by swarms of decoys, such as balloons and metal foil, so the 'acquisition' technique is not so simple. However, if it does its job properly, the PAR fires off a large three-stage SPARTAN rocket (Fig. 13.21), which can intercept well above the atmosphere, at a height of 2–300 km, and destroy the enemy missiles by X-rays from a thermonuclear warhead.

But suppose the enemy missile gets through. Then waiting for it is SPRINT (Fig. 13.22), a very high acceleration two-stage rocket guided by a missile site radar (MSR) (Fig. 13.23), which can intercept at a range of about 40 km in the atmosphere. Each SPRINT battery (or missile 'farm' as it is quaintly termed) must therefore be close to the target that it defends, such as a great city.

All this costs money (Table 13.2). The basic SENTINEL system was supposed to cost between 5 and 10 billion dollars but most people thought that this was likely to be only the beginning of an expenditure running up to 40 billion dollars. Notice, for example, the enormous cost of the radar systems – about 1000 times the cost of that primitive Chain Home line that served its purpose so well!

Table 13.2. *Cost estimates for ABM systems*

	Investment cost (billions of dollars)	
	Posture A	Posture B
Radar systems (PAR, MSR etc.)	6.5	12.6
Missiles (SPARTAN, SPRINT)	2.4	4.8
Defence Department	8.9	17.4
Atomic Energy Commission	1.0	2.0
Total investment cost (excluding R & D)	9.9	19.4
Annual operating cost	0.38	0.72
Number of cities defended	x	$2x$

For many people, of course, this vast expenditure was very welcome. Money for missiles eventually ends up as cash in the pockets of many thousands of people doing their honest jobs or taking their percentage as the goodies flow past. The scientists and engineers were fully involved on a very large scale. Research and development on ABM systems is said to have cost about $4 billion up to 1968, making this the largest technico/scientific effort ever undertaken. This expenditure could, some people say, have been justified even if the weapon was never put into the field, as a precaution against a breakthrough by the 'other side'. But it can also be argued that the major force for the design and construction of such systems is not military demand but the pressure for innovation from the research laboratories themselves. They are being paid to think of new weapons and they are unhappy and restless if their brilliant ideas are not used.

This pressure can defeat itself. The backroom boys are also employed

13.24 The 'fireball' of a nuclear explosion in the upper atmosphere blacking out radar observation of attacking missiles.

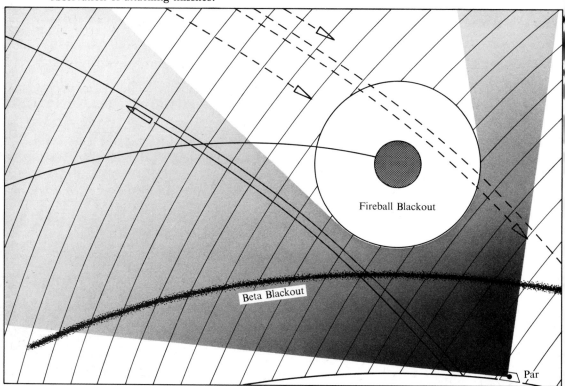

Fireball Blackout

Beta Blackout

Par

to think of ways of frustrating their own inventions. The ABM radar system is, in a fairly obvious way, susceptible to all manner of decoys, jamming with 'chaff' (i.e. clouds of fine wires), or blacking out by a nuclear 'fire-ball' of ionized particles (Fig. 13.24). Presumably the Russians are just as clever at the same game and could play other tricks that nobody has thought of. The uncertainties of the weapon–counter-weapon game in peacetime make the scientists seem even more important than they are in actual war, when their marvellous devices are tested in action against a real enemy.

Indeed, in December 1967 a new set of initials turned up – MIRV – Multiple Independently-targeted Re-entry Vehicle (Fig. 13.25). Each main rocket contains several nuclear warheads, each of

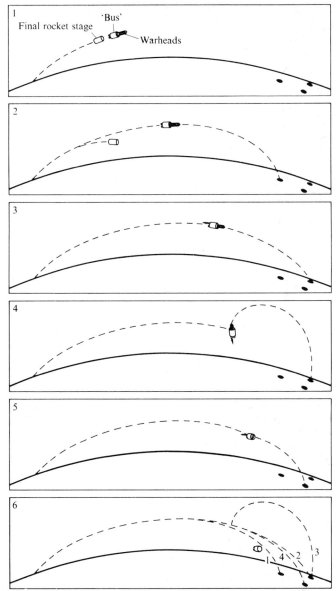

13.25 MIRV
(Multiple Independent
Re-entry Vehicle).

13.26 Scientists demonstrate against ABM.

which can be detached separately to hit a different target. The military effect of this could be to nullify the US riposte after a surprise attack from (say) the Soviet Union, since these warheads could be aimed to destroy American missiles in their silos before they could even be launched.

Whatever may have been happening in the back rooms, the installation of the SENTINEL system began in 1968. But when local people saw that ABM missile farms were being set up in the suburbs of their own cities, a great political rumpus was raised. Nobody wanted nuclear warheads capable of being fired at a few seconds notice, in their own backyards, even for their own protection. So the new Nixon Administration switched quickly to a new system called SAFEGUARD: the ABMs were to be set up around the offensive missile silos, away out in the wilderness, to protect them against a surprise first-strike attack.

In the summer of 1969 a big political debate broke out in Congress over the future of the ABM system. For the first time the scientific and technical opposition to anti-missile defence came out into the open, and made a strong public case in articles, books and expert testimony before congressional committees. Public protests, such as a march of 100

328

physicists to the White House (Fig. 13.26), probably had much less influence than the opinions of very distinguished experts – including five former Presidential science advisors and other top-level 'authorities' – concerning the likely capabilities of the weapon and its effects on the strategic balance of power. But one must be sorry for the poor senators having to decide between conflicting estimates of the likely attainable accuracy of hypothetical Russian MIRVs, and of the number of US Minuteman missiles that would be left after the most efficient likely attack. The differences between Professor Albert WOHLSTETTER (pro ABM) and Professor George W. RATHJENS (con ABM), for example, depend upon assumptions concerning the blast resistance of a Minuteman silo; would it be knocked out by an explosion of strength 1 megaton at a distance of ⅛ mile or not (more rigorously: 'one can either: (a) use the Packard chart and a CEP assumption to read a PK, or (b) use the Nitze data and an exponent for scaling lethal radius with yield, to deduce PIs for 1-MT using the PKs for 50-KT – but not both').

By this time, the whole analysis of whether it is better/cheaper/wiser etc. to have MIRVs and/or ABMs had become hopelessly confused. Weapons analysts invented parameters such as the

$$\text{'cost exchange ratio'} = \frac{\text{dollar cost of adding missiles to make up for those 'taken out' by ABMs}}{\text{cost of ABMs}}$$

But this is uncertain by a factor of three, which could make quite a difference to the human outcome as these figures show (Table 13.3). These quantitative arguments are no more convincing than verbal logic when the data are so vague.

Table 13.3

US fatalities (in millions)	Cost exchange ratio
40	1/4
60	1/2
90	1/1
100	Undefended case

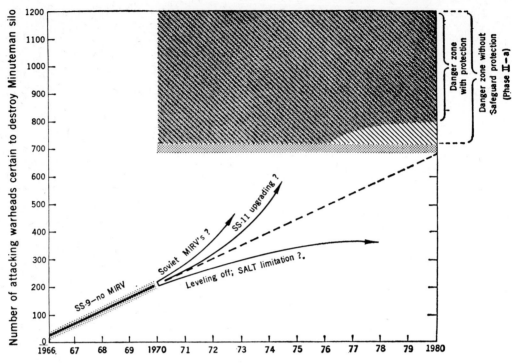

13.27   Graph of possible Soviet moves in weapon development.

In the end, the opponents of ABM lost by only one vote in Congress. In 1970 the Defence Department got their way more easily but what with the Strategic Arms Limitation Treaty (SALT) and various diplomatic détentes between the superpowers the ABM issue has changed in significance (Fig. 13.27). The Russians, by the way, gave up their ABM system for Leningrad, which sparked off the American reaction, and have only a thin defence for Moscow. The fantastic, monstrous 'deterrent' weapons systems still exist as a desperate threat to all mankind, but the political factors assumed in many analyses have altered in weight and form. We can only suffer, and comment on, history as it makes itself or is made.

THE TECHNICAL SOLDIERS

But the scientists are now completely integrated into the war-making machinery of the state. The temporary mobilization of the scientific community in the World Wars has been made permanent. The advanced

330

nations are spending a significant fraction of their national incomes on R & D for direct military purposes – the UK, for example, operates the most research-intensive military industry in the world, spending (in 1964), £62.2 for every £100 of military equipment bought (see p. 251). Very large numbers of scientifically trained men and women are occupied full time with the conception, design, production and testing of weapons for killing other men and women or for destroying the fruits of their labours.

This is not a marginal phenomenon of no more than symbolic significance. It means that something like one quarter of the students being taught science or engineering are going to be doing this work (Fig. 13.28). It means that a great many of our leading scientific authorities – not merely the administrative Big Chiefs but the intellectual Medicine Men as well – are personally involved, even in peacetime, with decisions on military technique and strategy. Men like Hans BETHE, Freeman DYSON, Paul DOTY or George KISTIAKOWSKY, in the class of Nobel Laureates, are respected for their contributions to pure knowledge, not for their political power: yet they find themselves acting as potentates, forced to throw their influence on one side or another, privately or in public, as consultants, advisors, witnesses, *experts* in this monstrous perversion of social needs. Neither they, nor their Soviet, Chinese, British, French, Japanese or Israeli opposite numbers are to be blamed; this is the world in which we all live.

We have even witnessed the birth of a new scientific specialty directly concerned with war. The radar, rocket and nuclear scientists are now joined by the 'strategic analysts' who tell the soldiers how they should really fight their wars. These are the successors of the operational research specialists of the Second World War, who showed by simple statistical analysis how best to arrange servicing schedules for aircraft, or what was the optimum depth to fire a depth charge dropped near a submarine. In those innocent days they applied their mathematics to problems of tactics and logistics: throughout the Cold War they have been fighting hot strategic scenarios on their computers (Fig. 13.29) looking for game-theory solutions in which not all the combatants were left in ruins!

This is an interesting intellectual development – an attempt to make a 'science' out of conflict and war. War has, of course, always had its

13.28   Advertisement for military research job.

13.29 Strategic analysis.

expert practitioners, its professionals, its training system, its colleges, its universities, like any other technology. There have been famous theorists such as CLAUSEWITZ (1780–1831) who came close to a philosophical turn of mind. But in the RAND corporation and the Hudson Institute – the so-called 'Think Tanks' supported by the American military machine – the attempt has been made to carry this science to the final quantitative stage. If one is sceptical as to the value of the advice they give on essentially unquantifiable subjects ('How many begabucks

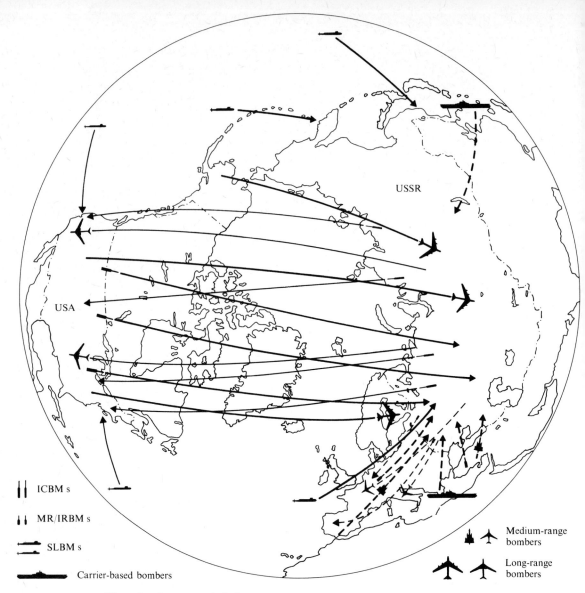

ICBM s

MR/IRBM s

SLBM s

Carrier-based bombers

USSR

USA

Medium-range
bombers

Long-range
bombers

13.30 Map showing strategic balance.

per megadeath?') one can only be fascinated at the intellectual virtuosity
thus displayed in such a bizarre mission.

The sad thing that has to be said is that most scientists employed in
military research are quite complacent about it. They justify themselves
by an appeal to their patriotic duty, and do their technical jobs with
complete devotion. This is not a criticism of their moral judgement: it is
simply a statement of fact.

Nevertheless, a tradition of dissent and counter-argument by experts in the name of more universal human values has been created in the past twenty years – since Hiroshima, perhaps – and is apparently growing in strength. It was something new when such scientific mandarins as KILLIAN, KISTIAKOWSKY, WIESNER, HORNIG, RATHJENS, YORK, BETHE, GOLDBERGER, PANOFSKY and DRELL shed the cloak of private influence in Washington, and spoke in the public political arena against ABM – not only as 'concerned citizens' but as real experts on nuclear warfare. The campaign against chemical and biological weapons, and against some of the worst technical excesses in the Vietnam War, is in the same tradition. Big Science has grown to maturity in the United States, and is now learning the burdens and moral responsibilities of being an Estate of the Nation.

At the heart of these movements is a strong anti-war sentiment amongst scientists, based upon personal experience of international cooperation (p. 225) and upon the moral conviction that science ought to be for the general good of mankind. The universality of scientific knowledge, the strenuous search for a consensus that transcends political frontiers, is completely antithetic to militarism and aggressive nationalism. These fundamental features of the scientific life are well understood by most leading scientists, and the sentiments to which they give rise are sincerely held, even if they are sometimes corrupted by political ideologies.

One should, for example, take notice of the Pugwash movement which has tried from the first to take a positive international line and to open up human communications, to clear up misunderstandings, and to devise techniques that would encourage mutual trust and a favourable climate for disarmament. This has been a special task of a portion of the scientific élite of the world: nobody yet knows (or perhaps will ever be able to judge) how much it has done, but it was surely the right thing to do. War also has its genuine opponents amongst scientists as amongst other men.

14 SCIENCE AND SOCIAL NEED

As long as we do not pretend to be competent, no harm is done.   *Arne Naess*

Let us now try to look at science from the outside – that is, from the viewpoint of the 95% or 98% of the population who are not scientists, technologists, engineers, or physicians. What does science mean for them? Why should they support it so lavishly with their taxes?

The most obvious individual need has always been for material well being. At all times, in all places, men have struggled to provide themselves with food, shelter, transport, and other goods and services. To fulfil such needs has always been one of the prime purposes of science. Go back to Francis BACON or to the manifestos of the founders of the Royal Society, and you will find a boundless faith in 'The Improvement of the Manual Arts' that was to come by observation, theory, and experiment. The Presidential Address to the British Association by W. FAIRBAIRN in 1861 is typical of the attitude of the thoughtful and sincere mid-Victorian man of science.

The history of man throughout the gradations and changes which he undergoes in advancing from a primitive barbarism to a state of civilization shows that he has been chiefly stimulated to the cultivation of science and the development of his inventive powers by the urgent necessity of providing for his wants and securing his safety.

The modern demand that science be useful and socially responsible is merely a re-emphasis of ancient virtues that have always been taken for granted.

Science has, of course, fulfilled almost to satiation these material needs. It is unnecessary to catalogue its medical, agricultural and engineering achievements. Through these we live quite different lives, in a different world from that enjoyed and suffered by our own grandparents. The fact that this transformation of the human condition can give rise to further needs and other serious problems is a separate issue, to which we shall later return.

At the same time we must not neglect the spiritual needs to which science caters. Human curiosity about the world in which we live is not an unreal psychological force. The need to find out about things should not be underestimated or discounted; this is a peculiar error that is often made by those who deplore the 'materialism' of scientific knowledge and its practical consequences.

SCIENCE POLICY

Until recently, science has responded to these needs haphazardly, by spontaneous growth within a general policy of *laisser faire*. Research has followed the inclinations of individuals or has swayed one way or another in response to fashion, or to the consensus of the privileged groups who wield scientific authority.

In practice, this means that decisions are taken in accordance with Alvin WEINBERG's *Internal Criteria of Scientific Choice*. Before embarking upon a research programme we tacitly ask questions such as: is the particular topic ripe for exploitation; are there good things to be done in that field; is the subject not too stale and overworked; are good people available to do the research? These questions are all directed towards the achievement of progress *inside* the particular branch of science, almost as if for its own sake. The advantage is, of course, that they can often be answered fairly accurately by the scientific experts in that branch, who at least guarantee that effort will not be wasted in fruitless or trivial endeavours. If our aim is simply to 'push back the frontiers of knowledge' by 'pure', 'basic', 'fundamental' research then they are by no means inappropriate criteria. In fact this has usually been the strategy implicit in the 'peer review' method of allotting funds through research councils, the National Science Foundation, and similar organizations (p. 246).

But the world-wide community of science is too large, and too closely connected with practical affairs, for such a policy to continue. A rigorous imposition of the internal criteria would not make the best use of the large numbers of aspiring, technically trained, research workers that are

337

now clamouring for employment. The amount of money available for fundamental research is limited, and would be spread too thinly if everybody were to be supported in research of his own choosing. Serious political and economic thought, and the experience of success in military and industrial research, clearly indicate that science should be planned to achieve much more explicit goals *outside* the scientific community.

In the nature of things, research results cannot be planned in detail. Who can know the outcome of a new experiment? If the result could be predicted, then the experiment would be unnecessary. To draw up a precise programme of investigation is likely to waste effort, and can even be self-defeating: the attempt to carry out a preconceived programme merely shackles the imagination in its drive to see things from a new point of view. There is an inner logic of nature that prevents premature success. For example, a determined effort to build a heavier-than-air flying machine would have been fruitless before the invention of the internal combustion engine. And an enlightened project, in 1850, to improve the horse as a means of transport would surely have failed to invent the automobile! The jargon of industrial production – 'norms', 'percentage fulfilment of the Plan' etc. – are ludicrously inappropriate to scientific activity.

But the objections to planned research can be overstated. Within 'normal' science, the range of problems worth tackling can be defined, and the general prospects of a successful outcome can be assessed. Just as in technological development, progress in the exploration of a newly opened field of basic research depends as much on a succession of small, relatively predictable steps as on single big jumps of imagination. The enormous numbers of scientists now at work in the world are not all in LANDAU's first class of ability (p. 122). Their capacities may well be put to best use in routine tasks within an organized scheme. A method whereby financial support is allocated according to some rough general plan may achieve an overall balance of effort without great constraints on individual scientists.

This is what is now called 'Science Policy'. The task of the policy makers, whoever they may be, is to observe gaps in the research front, or to take account of important social needs such as new sources of energy or the protection of the natural environment, and to manoeuvre squads of scientists into the appropriate direction of attack.

In this task it is natural to apply WEINBERG's *External Criteria for Scientific Choice*. We might assess a project, in the first instance, for *Technological Merit*: does the research lead to fairly obvious improvements of existing or proposed techniques? This would apply, for example, to careful measurements of the fission cross section of atomic nuclei that would be required in the design of a breeder reactor. We might look for *Scientific Merit*: does development in the field have important consequences in other fields? Thus, molecular biology has enormous long-term significance for all branches of medicine and biology; whereas discoveries in high energy physics, however beautiful in themselves, are not likely to have much impact on solid state physics or chemistry. Or we might apply the criterion of *Social Merit*: does this research have potential applications of great social value? By this criterion pure genetics is always significant since it touches on man in a fundamental aspect, and is thus capable of producing exceedingly powerful new techniques that could be of great human benefit.

THE ORGANIZATION OF SCIENCE

Merely to state these criteria does not tell us how to weigh them against each other. In the end, science policy depends on value judgements – the assessment of likely success, imagination of possible consequences etc. It has become one of the higher arts of government, with its own skilled experts, its own technical literature, its own triumphs and failures.

Much of the discussion revolves around the pattern of organization of research – the chains of command, information gathering and budgetary control that link the top level policy makers with research institutes, universities and individual scientists. To a large extent, these administrative procedures depend upon the style of government or industry within which they are embedded. In France the scheme is rational and centralized, like the whole French government machine: in Britain it is empirical, opportunist and greatly dependent on the personal links between the top people in various spheres of influence: in the United States it is pluralistic and competitive, reflecting the division of powers between various semi-autonomous agencies and corporations. There is no ideal formula for the construction of a research machine for all countries and for all seasons.

Consider, for example, the question whether there should be a *Ministry of Science*, comparable with a Ministry of Defence or of Education. The arguments for this are simply stated. Science has a certain unity of method and of techniques, and is practised by men and women of similar qualifications and style of life. As we have seen (p. 264), research can best be considered as a large-scale capital investment by society, uncertain in its immediate returns but promising very large returns in the long run. The production of knowledge is what the economists call an *indivisible communal commodity*, which is not easily adapted to the forces of the market place. Capitalist incentives of profit are not satisfactory, since they can lead to waste of resources, profiteering, and emphasis on trivial or harmful innovations. Society as a whole should determine its own ends and its own priorities: the planning of science becomes part of the general governmental responsibility for social foresight and wise investment. A voice for science at the Cabinet level thus seems essential.

But if we think about an applied science, we presume that it is directed towards particular ends. Research on nuclear energy, for example, is part of the 'mission' of those who build power stations: research on human disease is obviously related to the work of hospitals and health services. It makes good sense to assign responsibility – and expenditure – for such research to the organization governing each such activity. Research on transport systems should be sponsored and paid for by the transport industry, or by the Ministry of Transport, and not be counted as one of the functions of a general Ministry of Science supported by a 'Science Budget'. From this point of view, we may observe the relative variations of research effort on various social activities as symptoms of the differences of emphasis placed on these activities themselves: the poverty of research and development in the railway industry (p. 254) is merely a by-product of lack of investment in railway transport. In other words, the task of the scientific policy makers is not to provide research results for application here and there, but to ensure that a scientific attitude to innovation, and adequate research facilities, are made available for use in every significant activity of society.

The social function of *basic* or *pure* research is much more complex. We can think about it in three different ways, each of which suggest a different mechanism of administration and financial support.

1. *Basic research is the necessary back-up for applied research, although more speculative and requiring a longer term to bear practical fruit.* From this point of view, a pure, academic science, such as the quantum theory of the metallic state, is justifiable because it could, in twenty or fifty years time, form the basis for new practices in the manufacture of alloys. We have noted a number of cases (chapters 2, 7, 8) where fundamental scientific principles have eventually proved immensely valuable in the creation of new technologies. Although we may not see such uses now, it takes little imagination to believe that a high proportion of our present pure sciences are 'potentially applicable' in this sense. The support of basic science is thus a proper charge on all organs of applied research and development, to the extent perhaps of something like ten per cent of the money and men employed. In other words, we should follow the example of the Bell Telephone Company and build teams of pure scientists into every industrial research laboratory.

2. *Pure science is a great aesthetic and spiritual adventure for mankind, worth doing for its own sake like any art form.* What is our affluent and sophisticated civilization supposed to be doing, if not producing this and other forms of humanly valued product? Astrophysics, and the theory of elementary particles, may never produce much bread and butter but to get some glimpse, through them, of the nature of the universe is worth more than any material benefit. This activity should therefore be supported by society as a general good, in much the same spirit as we pay for public parks and symphony orchestras. The fact that some of this research is very expensive (chapter 9) and can only be done by very large teams of full-time professional scientists, does not excuse us from doing our spiritual duty by paying handsomely for it.

3. *Academic research is valuable because it is what academics are inclined to do.* Education in the higher scientific disciplines is necessary in modern society to produce competent technical experts. Those who teach such disciplines can only keep their minds sharp, and their science in good critical trim, if they are actively engaged in the intellectual competition of

research. From this point of view it matters little what problems they attempt to solve provided they work hard to solve them. Modest support of academic research is not directly 'productive', but in a roundabout way it has great value because it concentrates the best intellects in the universities and concentrates their minds wonderfully. Pure science thus becomes the responsibility of the institutions of higher education and research prowess is to be regarded as the most important attribute of the university teacher.

These points of view provide the extreme positions of a triangular debate within which we may find most of the various rationalizations of support for basic science. There is no standard recipe for the most efficient mixture of the utilitarian, romantic and academic motives for research: it depends, in each case, on the local conditions and priorities. But the gravest folly, exemplified in many developing countries during the past two decades (chapter 11) is to treat 'pure science' as a single uniform activity, to be supported at all costs because it is the modern thing to do or because it promises a short cut to wealth. Where resources are limited, the application of the criteria of scientific choice must be particularly careful and searching. The final tragedy is not that many wrong decisions were taken, leading to terrible extravagances on 'prestige' research projects, but that there may now be a reaction against basic science in favour of very short-term technology.

Of course the distinction between 'pure' and 'applied' science is arbitrary. A modern science-based technology, such as solid state electronics or the manufacture of pharmaceutical drugs, is intimately connected with academic, fundamental research, from which new discoveries, new techniques and new theoretical discoveries are continually being drawn. The existence of this market for research findings stimulates scientific activity in these fields and makes sure that plenty of money goes into the appropriate basic research. There is little practical difference between, say, the attempt to understand the mode of action of an existing antibiotic and the development of an improved formulation of the same drug.

According to the mythology of science-based innovation, the pure scientists should always be on the watch for fruitful applications of their research: in practice this is not so easy (see chapter 8): it often turns out that scientists who are hired to do applied research are so seduced into open-ended basic research that they forget what they are supposed to be doing. It was recently observed, for example, that in the study of magnetic materials – a field with many applications such as tape recording, ferrite aerials etc. – about 60% of the published papers were on fundamental topics, which often failed to deliver any goods. This trend can be noticed especially in large government and industrial laboratories (p. 199), such as those supposed to be devoted to the development of nuclear power: in the absence of strong market forces or decisive leadership, these can lose their sense of technical 'mission' and become mere confederations of independent research groups studying essentially academic topics. Many of the claims for support for 'socially relevant' research are thoroughly fudged by this method.

WHO MANAGES SCIENCE?

The really difficult problem in any national policy for science is to notice, or to anticipate, social needs, and to judge whether technical possibilities exist by which they might be met. The machinery by which such needs are made explicit must depend on the political and economic framework of each country: but whether these come as newspaper articles, decisions of the Party secretariat, questions in Parliament, or company prospectuses, the question 'can it be done?' must eventually pass into the hands of scientific experts. To whom can such important technical decisions be safely entrusted? It is natural to turn to the 'authorities' – the Academicians, the Royal Society, the directors of research institutes, the professors – for this advice. But the nominal leaders of the scientific community may be too old (see Fig. 3.34), too taken up with ceremonial duties, to be in touch with the latest developments. Or they may be heavily committed to out-of-date disciplines or to tasks that have become routine. A field of science that was once dynamic and progressive can go stale and lose its intellectual purpose, and yet continue to be pursued out of habit and early training. First class scholars often have an astonishing capacity to retain their intellectual vitality into old age; but the balance of disciplines in an Academy of scientific elder statesmen is not likely to represent the reality of the forces at work in present-day laboratories. *Gerontocracy* – the rule of old men – is a very real danger in a rigid and hierarchical academic community.

343

For this reason, the prestige rank order within the republic of science is an unsatisfactory basis for an administrative structure or for the making of policy. As we saw in chapter 6, successful scientific administrators are not always the most creative research workers of their day. It is one thing to win a Nobel Prize for brilliant intuitions about three-dimensional chemical molecules: it is quite another matter to manage a research institute with tact and energy or to assess the potentialities of polymeric materials in industrialized building. It is particularly important that attention be given to younger men, to critics of the established order, to imaginative outsiders and non-scientists, to the voices of novelty and unorthodoxy. It is the responsibility of the scientific community to the powers-that-be to present a sympathetic survey of the whole range of potentialities in a technological or scientific development rather than a cautious reaffirmation of the conventional wisdom. The good scientist is trained against putting on paper any opinion that he cannot substantiate by experiment or theoretical deduction: this virtue can blinker his view of the unpredictable future in which policy must act.

THE ASSESSMENT OF TECHNOLOGY

The applied scientist or technologist has another weakness as a policy-maker. He yearns for a '*Technological Fix*'. Asiatic poverty is to be banished by electrification, tractors, fertilizer factories and nuclear desalination plants: the transport problems of the cities will be solved with automatic hovertaxis: computerized teaching machines will make all children equal – and so on. Even in the field of medicine, with its long, long history of intimate interaction between technique and practice, there remains the belief that all human suffering will be banished by the medicaments and mechanical appliances to be invented by medical research, and inadequate attention is given to social and psychological factors.

Along with this goes the game of *technological forecasting*, where one guesses at all sorts of desirable and/or undesirable tricks and devices whose imminent introduction is then predicted to be inevitable. As a more disciplined, less romantic variety of science fiction, this game has great value in awakening the imagination and drawing attention to possibilities and trends that might otherwise be overlooked; but our experience during the past century has shown that such forecasts become completely unreliable over more than a few decades. The most successful 'prophecies' have come, not from technical experts, but from men of letters – H. G. WELLS' moon rocket, Olaf STAPLEDON's atomic weapons, Aldous HUXLEY's contraceptives. Observe, incidentally, that we are not all buzzing about in our own private helicopters, as was confidently predicted in the 1930s, but are now finding it more convenient to go by train 'as God intended us to do!'

The objection to the technological fix is that few of the needs of society can be met by such narrowly conceived and crudely programmed innovations. The great achievements of concerted, large-scale, mission-oriented research have been *technical*: discover an enemy aircraft, design a vaccine against polio, get a man on the moon. But many of the problems of society are, in Jerry Ravetz' term, 'practical'; they are the consequences of historical circumstances, and are embodied in human beings with all their present prejudices, conflicts and immediate needs. The Vietnam War has demonstrated, in blood and terror, the impotence of extravagant mechanical technique against human willpower and cunning.

Recognizing the uncertainty of simple technical innovations, some *Futurologists* (aren't we all, up to a point?) put their faith in *systems analysis*, which is supposed to take account of the interaction terms between various components of the social mix. In designing a new transport system for a city, for example, we might introduce equations describing the effects of the proposed network on the places where people subsequently choose to work and to live, and then vary the parameters to simulate the effects of different pricing policies and tax regulations on the overall outcome. The optimization of such a system of equations is a good exercise for an electronic computer, and often shows up unforeseen consequences, but it only pushes the uncertainties of prediction a few years ahead. As we have seen (chapter 12), economic, social and psychological factors cannot be represented with any accuracy by mathematical formulae and are often understood much better by shrewd politicians than by academic sociologists.

But this scepticism concerning the predictability of the consequences of technical innovation is not an argument for *laisser faire*. We are

345

all too familiar now with the emergence of fully-fledged, economically committed technical 'solutions' to problems that do not exist. The supersonic airliner, for example, can be seen as a product of sheer technical virtuosity, allowed to grow, by neglect of policy, as a result of general overall support of aircraft development and research, until it seemed essential actually to make it. In the field of civil technology, just as in military R & D, the mere possibility of an ingenious innovation becomes an important argument for putting it into practice, regardless of market demand or social consequences. It is certainly wiser to attempt to canalize these intellectual forces and human skills towards obviously desirable ends – quieter, cleaner automobiles and faster, steadier trains – even if we cannot be certain of success, and may accidentally introduce unwanted side effects which will cost us yet more money and research effort. The arrogant assertion that 'anything that can be done will be done' was intelligible only in the very special historical context of the United States (or the Soviet Union?) of the 1960s, when it almost seemed as if the R & D tail could wag the dog of social need. Crude technocracy is not in favour these days.

SCEPTICISM

The historical growth of a scientific approach to the practical arts, discussed in chapter 7, was often very slow and halting. Nevertheless, professional etiquette amongst the technical experts may encourage claims to special skills, based on high-faluting theories and 'a life-time of experience', that are not really justified. Medicine provides many good examples (p. 162): for centuries patients submitted to being bled, for innumerable diseases, just because this was the treatment recommended by 'expert opinion'. It is very likely that the treatments we are advised to give to many practical problems in economics, sociology, politics, or psychology are equally foolish and useless. Not all the doctrines claiming the authority of science are in fact sufficiently mature to be trusted as guides to action. As we saw in chapter 12, scepticism of professional expertise is probably still the wisest policy towards most of the claims of the social and behavioural sciences.

This sceptical attitude is an important part of the role of science in society. The fact is that the behaviour of many people, and of many institutions, is often governed by beliefs that have no rational basis. Superstitions about food and sexual behaviour, medical fallacies, racial prejudices, economic and social doctrines, are widespread even amongst educated, civilized people. Folk Science (p. 301) is often difficult to combat precisely because it is usually a rationalization of some out-moded set of general principles given weight by the authority of a past generation of intellectuals, priests and scholars.

On these it may be that current science has nothing positive to say. I don't think we know whether it is healthier to drink water before than during a meal, or whether all boys pass through an Oedipal stage, or whether black men have stronger sexual appetites than whites. What society needs is the clear assertion of our ignorance on such matters, and the denial of any scientific basis for a superstition or prejudice. In the realm of general ideas, in the construction of a philosophical or ideological system, the role of science is not to bolster up this or that metaphysical principle but to proclaim the values of scepticism and open-mindedness on issues that cannot be settled by the type of rationality used in science itself.

Some people maintain (p. 299) that science itself is not free of 'values' and is in fact determined by the ideological atmosphere of the society in which it is created. In the large this is incontrovertible: we think as we are brought up to think, and cannot do otherwise. But such a comment, to be constructive, should lead on to an alternative system of scientific thought which should be markedly different – if not superior – to that which exists. Until this alternative is produced, and shown to satisfy the usual canons of logical consistency, potential falsifiability, predictive power etc., we may sleep easy on this score. In the social and behavioural sciences, of course, such alternatives abound; but since practically none of them pass all the tests of scientific credibility this need not disturb us. The natural sciences can certainly be wrong in detail on particular issues (e.g. Continental Drift! (p. 78)) but there is little substance to the opinion that the whole enterprise is thoroughly corrupt and only to be corrected by a new theory of politics or a new vision of eternity.

Many scientists are surprised and shocked by the anti-science attitudes that are now quite widely voiced. They are bewildered and personally affronted by violent attacks upon the virtue of their profession and upon their own individual integrity. But we must not forget that for the past century science has been treated with too much reverence. Modern cults of mysticism and irrationality are genuine reactions to the exaggerated claims of *scientism* – the naive doctrine that all human ills can be cured by generous doses of the 'scientific method'. This belief, which is implicit in the attitudes of many scientists and technologists, lies very deep in our contemporary culture. Like many semi-religious doctrines, the belief that benevolent rationalism, or psychological conditioning, or submission to the laws of social evolution, will create a heaven on earth has its attractions; but such simplifications are not characteristic of the complex world of nature revealed to us by scientific observation.

Much more significant is the revulsion against mere *technique*, which is so often materialistic and inhuman in its crudest forms. 'Run-away technology' is denounced as an attack on 'human' values: the jet aircraft speeds through the air at the expense of peace and quiet: the mechanical dexterity of the computer offends against privacy and careless freedom. It is logically wrong to blame 'science' for these offences: knowledge is a tool for action, but not the agent. It is perfectly correct, however, to argue that if there were no science then the human condition could not be changing so rapidly – perhaps in the long run for the worse. Conservation of the natural environment and of some of our hard-won comforts seems highly desirable, even if other benefits may be lost in the process.

Most people, however, know in their hearts that this battle cannot be won simply by turning off the knowledge machine. The impetus of technical changes is too great: even with the fundamental science that we now have, technological development could go on for half a century without faltering. Our current problems – especially the population explosion – would not simply go away if all the research labs and libraries were burnt to the ground. We cannot reverse the flow of history and return unharmed to the glorious age of Queen Victoria – or would it be of Periclean Athens? Political and economic action is feeble against technical change without the aid of science itself. The countervailing power to crude, inhuman technique is more sensitive, more human technology. If we make the appropriate scientific effort, we can quieten the jet engine, and design much more flexible programmes for bank computers and credit cards.

The call for 'social responsibility in science' does not therefore proclaim a simple solemn duty: for some it means that the scientist cannot do good in present-day society, and must turn to political action and revolution: for others it means an appeal to private conscience in judging what programmes of research should be pursued, in case the results might prove harmful: another interpretation is that the social ends of the employer of the scientist should be carefully scrutinized before the work is allowed to go on: or it may be simply a plea for great caution in the development of new technology, so that the defects of an innovation (witness thalidomide) may be detected before they become apparent in use: for many technical experts it means no more than that they must be continually alert to the potential dangers of their techniques, and never be too proud or too cowardly to 'blow the whistle' when something goes wrong. In the realm of science policy, the call is for priority for research on matters that directly affect the everyday lives of many people, rather than for the benefit of a few profitable or prestigious industries, or for war-making, or for the sophisticated intellectual delight of a few hundred high-brow academics. All these responsibilities lie on the shoulders of the truly conscientious modern scientist.

WHAT IT MEANS TO BE A SCIENTIST

Scientific research is still done by individuals, but is now much more closely integrated with other social processes. The scientist is no longer an outsider, permitted to indulge in his own personal hobby to suit himself: he has become a central agent in a wide range of social activities, as expert, adviser, innovator or even decision maker.

This *politicization* of science has significant effects on the scientist himself: it governs his upbringing and perturbs his inner life. Until recently, most scientists enjoyed research as 'the great game'. The goals were serious – the uncovering of the secrets of nature, for the good of man and/or to the glory of God – and the labour itself had its own rewards. To be immersed in a scientific problem is very much like trying for days, weeks or months on end to put together an enormous jig-saw puzzle, or having to solve an endless succession of difficult crosswords. Despite endless frustrations, the moments of insight are worth all the pains. Whether as a lowly technician or as a star performer, research is a deeply satisfying vocation for those with a taste for it (see chapter 6). This satisfaction remains; but it can be poisoned by the uneasy feeling that one is simply indulging oneself, rather than contributing to the great forward movement of the human intellect etc. It is difficult to retain the sense of a vocation when research becomes a 'responsible', 'dutiful' profession, serving the needs of recognizable clients and paymasters. The dedication of the scientist to transcendental goals preserved his innocence and guarded his personal integrity: as research becomes 'a job like any other', then the pursuit of excellence in actual scientific achievement is replaced by the drives of ambition, vanity and the exercise of power, both within the scientific community and into society at large.

It is easy enough to invent scientific problems that are agreeable and interesting to solve by one's own efforts: pure mathematics is that sort of game. But to achieve planned social ends, it is necessary to undertake team research in the Big Science style. This is a very different sort of life from the academic ideal, and much more amenable to the manipulative skills and petty corruptions of the 'organization man'. Only under exceptional circumstances, such as at TRE (p. 320) or at Los Alamos (p. 142) during the War, is it possible to make a team of a big group of independent intellectuals. Quite different psychological types are needed to carry out such a complex, coordinated, scientific or technical effort as the Space Programme. The trouble is that the emphasis on reliability and *conformity* implicit in the bureaucratic structure of any such organization works against critical, innovative, self-confident *non-conformity*. The aim of a good Ph.D. course is to produce men and women with well-founded confidence in their own intellectual independence: the more science grows, the less room it may have for people with this quality of mind.

As the fundamental principles of the various branches of science are discovered and made clear, the need for continuous effort within the conventional disciplines is less evident. A much higher proportion of research work is likely to be concentrated on inter-disciplinary problems that are potentially applicable. This work will require a different type of scientist, able to play his part as a highly skilled specialist in a team but sufficiently broadly educated to understand the problem being tackled as a whole. This broader education, in which the social aspects of science will no longer be neglected, is also essential if he is to exercise his mind on the deployment or judgement of rational evidence in more general human contexts, as a citizen, as a teacher, or as a parent. The conflict with Folk Science begins at home: it cannot be left to general lectures and sermons by publicist professors: the way through the mess must eventually be found by many ordinary people trained in the scientific mode of thought and ready to say what they know.

Which really brings this book back to its beginning. This is the reason why a certain amount of exercise of mind and spirit on these difficult and multifold issues should be part of the education of every scientist – perhaps of every thinking person. Thank you for your attention!

QUESTIONS AND ANSWERS

The theme of this book is much too diverse to be covered by a finite list of sources and reference books. Most of the simple facts about people and events can be found in any good encyclopaedia, but to get to the interesting detail one must fossick around in a library, or search the contents lists of journals such as *Science, Minerva, Bulletin of the Atomic Scientists* etc. In any case, there is no great value in going very deeply into the particular topics which are discussed at length in the preceding pages: these are only examples of phenomena you can confirm by studying other similar cases for yourself.

The following lists of references are merely possible launching pads for such ventures and are in no sense complete.

It is easier to discover a lot of answers than to ask the correct questions. Just for fun, a few possible titles for provocative theoretical essays are also provided. Since nobody really knows beforehand on which side he should stand in many such controversial matters, I offer a choice of opposing starting points to suit all temperaments.

1 SCIENCE AS A SOCIAL INSTITUTION

Some very general references, relevant to most chapters:
The Sociology of Science edited by B. Barber & W. Hirsch (Free Press 1962)
The Social Function of Science by J. D. Bernal (MIT Press 1967)
Science in Modern Society by J. G. Crowther (Barrie & Jenkins 1968)
What is Science for? by B. Dixon (Collins 1973)
Science Observed by F. R. Jevons (Allen & Unwin 1973)
The Sociology of Science by R. K. Merton (University of Chicago Press 1973)
Scientific Knowledge and its Social Problems by J. R. Ravetz (Oxford University Press 1971)
Physics and its Fifth Dimension: Society by D. Schroeer (Addison–Wesley 1972)
Public Knowledge by J. M. Ziman (Cambridge University Press 1968)

2 WHICH CAME FIRST: SCIENCE OR TECHNOLOGY?

There are numerous histories of science, of varying reliability – for example:
Science since 1500 by H. T. Pledge (HMSO 1966) is a good elementary book
A History of Science, Technology & Philosophy by A. Wolf (2 volumes, Allen & Unwin, 3rd edn, 1962) covers the sixteenth, seventeenth and eighteenth centuries more comprehensively

Science in History by J. D. Bernal (C. A. Watts 1969) deals at length with more modern developments from a Marxist standpoint

Science for the Citizen by L. Hogben (Allen & Unwin, 4th edn, 1956) is readable and wide ranging

The History of Technology volumes i–v edited by C. Singer, E. J. Holmyard, A. R. Hall & T. I. Williams (Oxford University Press 1954–8) is the basic reference, of which

A Short History of Technology by Derry & Williams (Oxford University Press 1960) is an abridgement

Technology in Western Civilization edited by M. Kranzberg & C. W. Pursell (2 volumes, Oxford University Press 1967–8) is more readable

Science & Civilisation in China (many volumes, Cambridge University Press (1954–) by J. Needham is a mine of information for the real enthusiast

In addition, there are innumerable books about the history of particular branches of science and technology, as well as biographies and research papers. But beware of the historical notes in most scientific textbooks, which are usually grossly oversimplified.

Essay topics

Greek philosophy is irrelevant to modern science
Modern science was invented by the Greek philosophers
China had magnificent technology but no science
Chinese technology was science in action
Instrumentation – the missing link between science and technique
Telescopes and microscopes: toys of the idle rich
The conservation of energy and the economics of power
Electromagnetism – as useful as a new-born baby
Lysenko – the man who tried to do useful genetics
All scientific knowledge is potentially applicable
All science stems from technical need

3 WHO WAS A SCIENTIST?

This subject is not treated adequately, as a whole, in any one book:

The Social Relations of Science by J. G. Crowther (Dufour 1966) notes the main features

Men of Mathematics (2 volumes, Penguin 1953) by E. T. Bell is very interesting

Biographical Encyclopaedia of Science and Technology by I. Asimov (Allen & Unwin 1967) contains numerous potted biographies

Brief History of the Royal Society by E. N. da C. Andrade (Royal Society 1960) is a useful summary

History of the Royal Society by Thomas Sprat (1667; reproduced by Routledge & Kegan Paul 1959) is a famous early source

Science, Technology and Society in 17th Century England by R. K. Merton (Howard Fertig 1970) was an important step in the development of the new discipline 'The Sociology of Science'

The Scientific Intellectual by L. Feuer (Basic Books 1963) is provocatively
controversial

The Anatomy of a Scientific Institution. The Paris Academy of Sciences 1666–1803 by
R. Hahn (University of California Press 1971) is the best source

Science in Russian Culture by A. Vucinich (2 volumes: *History to 1860*, Peter Owen 1967;
1861–1917, Stanford University Press 1971) is authoritative

The Scientist's Role in Society by J. Ben-David (Prentice-Hall 1971) deals especially with
nineteenth-century Germany

The Organisation of Science in England by D. S. L. Cardwell (Heinemann 1972) deals
with the nineteenth century

Scientific Societies in the United States by R. S. Bates (Pergamon, 3rd edn, 1966) is a
standard source

Essay topics

Modern science is a direct product of puritanism
Modern science arose out of the bourgeois revolution
The Ph.D.: a fruitful German invention
The Ph.D.: a frightful German invention
The state as the patron of science
Science and national pride
Science and national prosperity before 1900
The Advancement of Science in the Victorian ethos
Whatever happened to the Royal Society?
Learned societies exist mainly for mutual admiration

4 STYLES OF RESEARCH

This must depend on good full-length biographies, such as

Michael Faraday by L. Pearce Williams (Chapman & Hall 1965), or autobiographical
material such as

The Double Helix by James Watson (Weidenfeld & Nicolson 1968):
The annual volumes of

Biographical Memoirs of Fellows of the Royal Society are full of revealing snippets
concerning less distinguished scientists of our own time

Science as a Vocation by Max Weber is a beautiful essay, to be found in most collections
of his writings

The Search by C. P. Snow (Macmillan 1958) is a good novel on the subject, although

The Struggles of Albert Woods by William Cooper (Penguin 1966) is more fun

Essay topics

Every scientist is, at heart, an amateur
'Fame is the spur', in science as in politics

A passion for research is a neurotic symptom
Originality is not enough
The best scientists are necessarily political radicals
Science is an inhuman occupation

5 SCIENTIFIC COMMUNICATION

Public Knowledge by J. M. Ziman (Cambridge University Press 1968) presents my own
  views on this subject
The Scientific Community by W. O. Hagstrom (Basic Books 1965), like many other
  books on the sociology of science, describes various practices and defects of the
  system
Introduction to Information Science edited by T. Saracevic (Bowker 1970) is a real
  technical treatise
The Social Function of Science by J. D. Bernal (MIT Press 1967) proposed a new system
Scientific & Technical Communication is a report (1969) of the U.S. National Academy of
  Sciences – and there are many more
Science since Babylon (Yale University Press 1961) and *Little Science: Big Science*
  (Columbia University Press 1963) by D. J. de S. Price assesses the growth of the
  scientific literature
Science: Growth and Change by H. W. Menard (Harvard University Press 1972) is full
  of provocative quantitative evidence about one field of science
The Structure of Scientific Revolutions by T. S. Kuhn (University of Chicago Press
  1970) expounds an important thesis
Personal Knowledge by M. Polanyi (Routledge 1962) and
The Logic of Scientific Discovery by K. Popper (Hutchinson 1968) are basic texts
  relating the philosophy of science to the communication of knowledge

Essay topics

The Invisible Colleges of Science
Mutual criticism is the cornerstone of the scientific method
Broken English – the essential universal language of science
Text books and paradigms
The scientific paper, and what has become of it
Automated information dissemination
How does information become knowledge?
Publish and/or perish!
Scientific conferences – the ideal form of conspicuous travel
Science can never be popularized
All science must eventually be popularized

This subject has scarcely been explored at all, although it is obviously relevant to the sociology of the scientific community and to the vast literature on the influence of science on general philosophy and social ideology – for example:

Science and the Social Order by B. Barber (Collier Macmillan 1962)

Science and the Modern World by A. N. Whitehead (Macmillan 1926)

Where the Wasteland Ends by T. Roszak (Faber 1973)

The most useful works are biographies of the leading figures of each age, and contemporary comments on their personalities and consequences.

Politics and the Community of Science by J. Haberer (Van Nostrand Reinhold 1971) is sobering

The Politics of American Science by D. S. Greenberg (Penguin 1969) is shocking

The role of authority *within* the scientific community is discussed in:

Social Identification in Science by J. R. & S. Cole (University of Chicago Press 1973)

Essay topics

Science is democratic in principle, but élitist in practice

Science is élitist in theory, but democratic in reality

Newton – the hero of British science

Einstein and Nazi Germany

How to win a Nobel Prize

If you can't do research, at least you can teach science

If you can't do research, you can't teach science

Philosopher kings, and the power that corrupts

Good scientists are seldom good scientific advisers

Only good scientists should be given political power

7 FROM CRAFT TO SCIENCE

See the notes on chapter 2. Also, various histories of various institutions such as universities, hospitals, professional guilds. Typical general work include:

The History of Medicine by B. Inglis (Weidenfeld & Nicolson 1965)

The History of Medicine by D. Guthrie (Nelson 1945)

A Social History of Engineering by W. H. G. Armytage (Faber, 3rd edn, 1970)

Technology & the Academics by E. Ashby (Macmillan 1958) is shrewd on technological education

Technics and Civilization by C. Mumford (1934; Routledge 1946) is the sort of 'philosophical' work against which the more specialist material should be set

Essay topics

The distinction between science and technology is meaningless
Science and technology serve quite different ends
Medicine is more a science than an art
Medicine is more an art than a science
A well-trained brewer is a well-educated man
Fundamental pure science is irrelevant in the education of an engineer or doctor
The only sure foundation of technological education is pure science
All techniques will eventually be made scientific
No technique is truly scientific
Technology is essentially conservative
A scientific technology cannot stand still

8 INVENTION, RESEARCH AND INDUSTRIAL INNOVATION

Relevant biographical works are, of course, essential primary sources, but the subject is well covered by:

The Sources of Invention by J. Jewkes, D. Sawers & R. Stillerman (Macmillan, 2nd edn, 1969)

Wealth from Knowledge by J. Langrish, M. Gibbons, W. G. Evans & F. R. Jevons (Macmillan 1972)

The Economics of Technological Change edited by N. Rosenberg (Penguin 1971)

Essay topics

The era of the backyard inventor is ended
Individual initiative is the motive power of innovation
Modern industrial innovation is the product of industrial research
The pure science of today has little effect on industrial innovation
The development phase of an invention
Scientists and patents
The suppression of inventions
The inventor or the research team
Technical knowhow and/or theoretical knowledge
Resistance to innovation
Inspiration, perspiration or scientific method
The transformation of a technique by the accumulation of minor inventions
Rational design of new products
The effect of the market on the sources of invention

9 BIG SCIENCE

This is a subject that has received little direct study. Most of the books on the management of research establishments tell us nothing.

Reflections on Big Science by A. M. Weinberg (MIT Press 1968) is an excellent collection of essays

Some aspects of life in high energy physics are studied in

Originality and Competition in Science by J. Gaston (Chicago University Press 1973)

Essay topics

Sealing wax and string
Instrumentation in modern science
The scientific instrument industry
Design criteria for big machines
Administering research
Planning a big, big experiment
Learning a technique or learning to do research
Team work in research
Leadership in Big Science
Multiple authorship and its consequences
The life cycle of a big laboratory
Funds and means in Big Science

10 PAYING FOR SCIENCE

This subject now has an immense literature – chiefly because most studies of the role of science in society eventually hinge on the amount of money politicians will spend on it and the mechanisms they devise to control the spending. In addition to most of the general books noted under chapter 1, this would apply also to:

Science and Society by H. & S. Rose (Allen Lane, Penguin, 1969)

The Scientific Estate by D. K. Price (Oxford University Press 1965)

The Politics of Science edited by W. R. Nelson (Oxford University Press 1968)

The Economics of Research and Technology by K. Norris & J. Vaizey (Allen & Unwin 1973) gives a modest synthesis of the economic arguments

The Economics of Technological Change edited by N. Rosenberg (Penguin 1971) and

Economics of Information and Knowledge edited by D. M. Lamberton (Penguin 1971) include various key papers

The statistical digests published by various governments, and the various reports by OECD are full of primary data

Essay topics

Space science: rewards and costs
Technological spin-off
How much science can we afford?
Grantmanship
The usefulness of useless research
Is duplication of research wasteful?
Scientific choice
Research as capital investment
Is science worth doing?
Does industrial research really pay?
Research, innovation and competition
Research expenditure in the Soviet Union

11   SCIENCE AS A CULTURAL IMPORT

The literature on this subject is very scattered. The task for the student is to collect information from people, newspapers, government reports etc. – or by personal observation.

Science Development: The Building of Science in Less Developed Countries by M. J. Moravcsik (International Development Research Centre, Indiana University 1975) admirably surveys the problems

The Brain Drain edited by W. Adams (Collier Macmillan 1968) discusses many of the key issues

Essay topics

Science outside the West
Colonial science
Exploration as a scientific activity
Science and cultural rebirth
The Agricultural Revolution
The influence of scientific knowledge on population growth
Science as a factor in economic development
A science policy for the Third World
Education and science in a developing country
Western science: the wrong road for a new nation
Science and socialism: the Russian experience
What is happening to science in China?

12   THE SCIENCE OF SOCIETY

There are, of course, great libraries full of relevant books. The following just happened to be useful in various ways.

The Behavioral and Social Sciences: Outlook and Needs US National Academy of Sciences & Social Science Research Council (1969). A somewhat optimistic account

Keynes and After by M. Stewart (Penguin 1970). Very neat summary

The American Soldier by S. A. Stouffer *et al.* (2 volumes, Princeton University Press 1942). Classic work

A Hundred Years of Sociology by G. D. Mitchell (Duckworth 1968)

Sociological Methods by N. K. Denzin (Butterworth 1971)

Sociology in Use by D. M. Valdes & D. G. Dean (Collier Macmillan 1965)

The Relevance of Sociology edited by Jack Douglas (Appleton-Century-Crofts 1970). Good critical perspective

Race Relations by M. Banton (Tavistock 1967)

Hereditary Genius and Inquiries into Human Faculty & its Development by Francis Galton (Macmillan 1869)

Intelligence: some recurring issues edited by L. E. Tyler (Van Nostrand Reinhold 1969)

Intelligence and ability edited by S. Wiseman (Penguin 1971)

The Appraisal of Intelligence by A. W. Heim (National Foundation for Educational Research 1970)

Secondary School Selection by P. E. Vernon (Methuen 1957)

Intelligence, Psychology and Education by B. Simon (Lawrence & Wishart 1971)

Inequality by C. Jencks *et al.* (Allen Lane 1972)

Essay topics

A science of society: our only hope

A science of society: a vain hope

A science of society: a terrible prospect

Why are natural scientists prejudiced against the social sciences

History as an exact science

Anthropology and its practical use

How scientific is economics?

Description in the social sciences

How to get truth out of statistics

Social indicators

Opinions as evidence

Intelligence and educational opportunity

The unity of mankind

The diversity of mankind

13   SCIENCE AND WAR

Surprisingly, despite all the controversy over the use of advanced technology in the Vietnam War, the subject has not been given adequate general treatment in historical or sociological depth.

The Challenge of War by G. Hartcup (David & Charles 1970) is a conventional account

Tongues of Conscience: War and the Scientist's Dilemna by R. W. Reid (Constable 1964) is a readable historical account, emphasizing various dramatic episodes

The Science of War and Peace by R. Clarke (Cape 1971) deals with the current situation

We all fall down by R. Clarke (Allen Lane 1968) is a good account of chemical and
biological warfare

Much of the best material is to be found in the *Bulletin of the Atomic Scientists*, which
has striven mightily over the past decades to keep the issues under serious discussion.

On Thermonuclear War by H. Kahn (Princeton University Press 1960) gives the full
flavour of the theoretical strategists

War: Studied from psychology, sociology, and anthropology edited by L. Bramson and
G. W. Goethals (Basic Books 1964) discusses some of the general psychological and
sociological causes of war

Essay topics

War is the crucible of science
Science is, by its nature, international and peaceable
Science is, by its nature, a source of violence
Scientists should never do military research
A scientist must always serve his country
Scientific spies
Modern scientific war
Scientific technology against humanity
Chemical warfare
Biological weapons
Swords and ploughshares: radar in war and peace
Nuclear weapons in war
The ABM controversy
The significance of J. Robert Oppenheimer
Defence support of pure science
Scientific consultants or technical soldiers?
The policy of strategic deterrence
Operational research
German science in the Second World War
Science in the First World War
The Pugwash movement: what has it achieved?

14 SCIENCE AND SOCIAL NEED

The general references (Chapter 1) would all be relevant. But one should also dip into
books such as:

Technopolis by N. Calder (MacGibbon & Kee 1969)

Science: the Glorious Entertainment by J. Barzun (Harper & Row 1964)

There is a lot now on technological forecasting; for example

Innovations: Scientific, Technological and Social by D. Gabor (Oxford University Press
1970)

The problems of science under political pressure are evident from:

The Rise and Fall of T. D. Lysenko by Zh. Medvedev (Columbia University Press 1969)
The Medvedev Papers by Zh. Medvedev (Macmillan 1971)
The First Circle by A. Solzhenitsyn (Collins 1968)
Politics and Philosophy in the Soviet Union by L. R. Graham (Allen Lane 1973)

Essay topics

The potential applicability of pure science
A plan for research
How can one plan to make a discovery?
The scientific Establishment
Gerontocracy in Science Policy
'Folk Science'
Pseudo-sciences
Scientism, an infantile disorder
Belief and scepticism in scientific research
Forecasting the future
A technological fix for Calcutta
A Hippocratic oath for scientists
Interdisciplinary research
Specialization in research and in scientific education
Why be a scientist?
Why not be a scientist?

PICTURE SOURCES

The author and publisher would like to thank the following for permission to reproduce illustrations.

2.1: BM Add. Ms. 23387 fol. 28*a*. *British Museum*. **2.2**: BM Or Ms. 15255 d.21. *British Museum*. **2.3**: S. R. K. Glanville, *Legacy of Egypt*, plate 22. Oxford University Press 1942. *Egyptian Museum, Cairo*. **2.4**: Dominique Halévy, *Histoire des Armes et des Soldats* (English translation, Odhams, n.d.). *Fernand Nathan, Éditeur*. **2.5**: *Science Museum, London*. **2.6**: J. W. Brailsford, *Guide to Antiquities of Roman Britain*, fig. 41. London, BM 1951. *British Museum*. **2.7**: BM Add. Ms. 18850, fol. 17v. *British Museum*. **2.8**: *British Museum*. **2.9**: *Mansell Collection, London*. **2.10**: C. Singer, *From Magic to Science*, fig. 70. Benn 1928. *Ernest Benn*. **2.11**: J. Needham, *Science and Civilisation in China*, vol. 1, fig. 29. Cambridge University Press 1954. **2.12**: *Herbal of Apuleius Barbarus,* formerly in Abbey of Bury St Edmunds. No. cxli, described by R. T. Gunther. Ms. Bodley 130. *Bodleian Library, Oxford*. **2.14**: Thomas Sprat, *History of the Royal Society*, 1667. Reprint ed. J. I. Cope and J. W. Jones. St Louis, Washington University Studies; London, Routledge 1959. *Washington University Press*. **2.15, 2.16**: *Correspondence of Isaac Newton*, ed H. W. Turnbull, vol. 1, plate iii, plate ii. Cambridge University Press 1959. *University Library, Cambridge*. **2.17**: *Istituto e Museo di Storia della Scienza, Florence*. **2.18**: *University of Cambridge, Whipple Museum of the History of Science*. **2.19**: *National Maritime Museum, London*. **2.20**: Singer *et al. History of Technology*, vol. 4, fig. 99. Oxford University Press. *ICI Ltd*. **2.21**: Data from Singer *et al. History of Technology*, vol. 4, p. 164. Oxford University Press. Engines redrawn from (*a*) Science Museum Neg. no. 7623; (*b*) drawing at *Royal Society*; (*c*) John Farey, *A treatise on the steam engine*, vol. 1, plate xi, 1827; (*d*) Science Museum Neg. no. 1252/73; (*e*) Singer *et al. History of Technology*, vol. 5, fig. 48. (*f*) *Engineer*, 26 January 1877; (*g*) *Engineer*, 1 May 1903. **2.22**: *Crown copyright, Science Museum, London*. **2.23**: Illustration from Louis Figuier, *Les Merveilles de la Science*, Paris *c.* 1870; in Bern Dibner, *Oersted*. Burndy 1961. *Burndy Library*. **2.24**: Singer *et al. History of Technology*, vol. 4, fig. 346. Oxford University Press. *ICI Ltd*. **2.25**: *Royal Institution of Great Britain*. **2.26**: Singer *et al. History of Technology*, vol. 5, fig. 76. Oxford University Press. *ICI Ltd*. **2.27**: *Engineer*, **67** (1889) 286. **2.28**: *Suttons Seeds Ltd*. **2.30**: *Iconographia Mendeliana*, fig. 64. *Moravske Museum, Brno*. **2.31**: H. P. Riley, *Introduction to Genetics and Cytogenetics*. Wiley 1948. *Professor H. P. Riley*.

3.1: *Royal Society of London*. **3.2**: *Cambridge University, Whipple Museum of the History of Science*. **3.3**: *Royal Society of London*. **3.7**: Photogravure of this portrait in *Oeuvres complètes de Christiaan Huygens*, vol. 1. Société Hollandaise des Sciences, La Haye 1888. *Cambridge University, Whipple Museum of the History of Science*. **3.8**: *Rijksmuseum voor de Geschiedenis der Natuurwetenschappen, Leiden*. **3.9**: *Cambridge University, Whipple Museum of the History of Science*. **3.10**: Based on a plate from S. Hoole, *The select works of Antony van Leeuwenhoek*, vol. 2, 1807: based on Leeuwenhoek's Letter 113 to the RS (12 January 1689). **3.11**: From Lampas, in R. T. Gunther, *Early Science in Oxford*, vol. 8, 1930. *A. E. Gunther*. **3.12**: T. Sprat, *History of the Royal Society*, 1667. Reprint ed. J. I. Cope & H. W. Jones. St Louis, Washington University

Studies; London, Routledge 1959. *Washington University Press.* **3.13:** *National Portrait Gallery.* **3.14:** ULC Ms. Add. 3958. 3, fol. 72r. *University Library, Cambridge.* **3.15:** *Royal Society of London.* **3.16:** *Neidersächsische Landesbibliothek, Hannover.* **3.17:** *National Maritime Museum, Greenwich.* **3.18:** *The Master and Fellows, Magdalene College, Cambridge.* **3.19:** *Academia dei Lincei.* **3.20:** *British Museum.* **3.21:** T. Sprat, History of the Royal Society, 1667. Reprint ed. J. I. Cope & H. W. Jones, p. 218. St Louis, Washington University Studies; London, Routledge 1959. *Washington University Press.* **3.22:** *Cliché Observatoire de Paris, J. Counl.* **3.23:** *National Portrait Gallery.* **3.24:** Detail of plate XI from *A Delineation of the Strata of England and Wales,* 1815. Photo *Robin Godwin, Department of Geology, Bristol University.* **3.25:** *Royal Society of London.* **3.26:** Derek J. de Solla Price, *Little Science, Big Science,* fig. 1, p. 9. Columbia 1963. *Derek J. de Solla Price.* **3.27:** *Royal Institution of Great Britain.* **3.28:** *Frank Wells.* **3.29:** D. S. L. Cardwell, *Organisation of Science in England,* p. 196. Heinemann 1957. *Heinemann.* **3.30:** Photo W. H. Hayles. From Lord Rayleigh, *Life of Sir J. J. Thomson.* Cambridge University Press 1942. **3.31:** *Robert Matthew, Johnson-Marshall & Partners.* **3.32:** Professor A. Delaunay, *Institut Pasteur, des origines à aujourd'hui.* Editions France-Empire, Paris 1962. *A. Delaunay.* **3.33:** *Royal Society of London.* **3.34:** M. Alfred Kastler, *C.R. Acad. Sci. Paris,* **276** (1973) 65. **3.35:** From a drawing by Edison's draughtsman (Samuel D. Mott). *Scientific American,* **41,** 16 (18 October 1879).

4.1: Windsor drawing 12281r. *By gracious permission of Her Majesty the Queen.* **4.2:** *Crown copyright, Science Museum, London.* **4.3:** Windsor drawing 12660v. *By gracious permission of Her Majesty the Queen.* **4.4:** *British Museum.* **4.5:** A. J. Berry, *Henry Cavendish,* fig. 6, p. 165. Hutchinson 1960. *Hutchinson Publishing Group.* **4.6:** *Phil. Trans.* 1798, tab. IV, p. 102. **4.7:** *Of the Propagation of Heat in Fluids: Collected Works of Count Rumford,* ed. Sanborn C. Brown, vol. 1, plate 1. *Belknap Press of the Harvard University Press.* **4.8:** *British Museum.* **4.9:** *Professor Sanborn C. Brown.* **4.10–4.13:** Alfred Wegener, *Origin of Continents & Oceans,* trans. Biram: fig. 4, p. 18; fig. 18, p. 72; fig. 30, p. 117; fig. 6, p. 28. Dover 1966. *Dover Publications Inc.* **4.14, 4.15:** James Watson, *The Double Helix,* p. 220 and facing p. 184. Weidenfeld & Nicolson 1968. *James Watson.* **4.16:** E. Segré, *Enrico Fermi, Physicist,* facing p. 117. Chicago 1970. *University of Chicago Press.* **4.17:** Enrico Fermi, *Notes on quantum mechanics,* p. 19. Phoenix Books 1961. *University of Chicago Press.* **4.18:** By arrangement with the *Trustees of the London Evening Standard.* **4.19:** Photo *Morning Star.*

5.1: *Correspondence of Isaac Newton,* ed. J. W. Turnbull, vol. 1, letter 48, plate IV. Cambridge University Press 1959. *University Library, Cambridge.* **5.2:** *Correspondence of Isaac Newton,* vol. 2, plate 1r. Cambridge University Press 1960. *British Museum.* **5.3:** James Watson, *The Double Helix,* pp. 227–33. Weidenfeld and Nicolson 1963. *James Watson.* **5.4:** *University Library, Cambridge.* **5.5:** *The Wellcome Trustees.* **5.6:** *Staatliche Museen Preussischer Kulturbesitz, Berlin.* **5.7:** *University of Cambridge, Whipple Museum of the History of Science.* **5.9, 5.10:** J. D. Watson & F. H. C. Crick, *Nature,* 25 April 1953. **5.11:** *Iconographia Mendeliana,* fig. 53. *Moravske Museum, Brno.* **5.12:** *Annalen der Chimie,* vol. XLII, 1842, p. 233. **5.13, 5.14:** D. J. de Solla Price, *Little Science, Big Science:* fig. 13, p. 44; fig. 1, p. 9. Columbia 1963. *Derek J. de Solla Price.* **5.15:** *American Institute of Physics.* **5.16:** *Royal Society.* **5.17:** *Instituts Internationaux de Physique et de Chimie:* founder, E. Solvay. **5.18:** *Institute of Physics, Bristol.* **5.19:** *Popular Scientific Recreations,* fig. 174 (translation of Gaston Tissandier, *Les Récréations Scientifiques*). London, Ward, Lock and Bowden, n.d. **5.20, 5.21:** *Royal Institution of Great Britain.* **5.22:** *Punch.*

6.1: *Histoire Générale des Sciences,* III-1, plate II, no. 3. (Originally in P. Dupuy, *La Vie d'Evariste Galois.*) *Presses Universitaires de France.* **6.2:** *University of Cambridge, Cavendish Laboratory.* **6.3:** Derek J. de Solla Price, *Little Science, Big Science,* fig. 13, p. 44. Columbia 1963. *Derek J. de Solla Price.* **6.4:** From Laura Fermi, *Atoms in the Family. Courtesy of Enrico Navone.* **6.5:** *Archives of the California Institute of Technology.* **6.7:** *Franklin D. Roosevelt Library.* **6.8:** *Royal College of Surgeons of England.* **6.9:** W. A. Shenstone, *Life of Justus von Liebig,* frontispiece. Cassell, 1895. **6.10:** Lithograph of P. Wagner from original by W. Trautschold & H. von Ritgen. *Liebig Museum, Geissen.* **6.11:** H. T. Pledge, *Science since 1500*, chart III, p. 200. HMSO 1966. *Crown copyright, by permission of the Controller of Her Majesty's Stationery Office.* **6.12:** Data from paper by H. Zuckerman, *Scientific American*, **217**, 5 (1967) 25. **6.13:** *Royal Society.* **6.14:** *National Portrait Gallery.* **6.15**, **6.16:** *British Museum.* **6.17:** Photo National Portrait Gallery. **6.18:** *Messrs T. & R. Annan, Glasgow.* **6.19:** *Crown copyright, by permission of the Controller of Her Majesty's Stationery Office.* **6.20:** *By courtesy of United Kingdom Atomic Energy Authority.* **6.21:** *United Press International (UK) Ltd.*

7.1: *University Library, Cambridge.* **7.2:** *World Health Organisation. Photo Paul Almasy.* **7.3:** Bibliothèque Nationale, Ms. gr. 2144, fol. 10v. *Hirmer Photoarchiv.* **7.4:** Osterreichische Nationalbibliothek, Cod. med. gr. 1, fol. 3v. *Osterreichische Nationalbibliothek.* **7.5:** *University Library, Cambridge.* **7.6:** Bologna University Ms. 3632 fol. 428v (Apollonius, *Dislocations*, chapter 1, in Greek). *The Wellcome Trustees.* **7.7:** Reproduced from *Medicine & the Artist (Ars Medica)*, by permission of *the Philadelphia Museum of Art.* **7.8:** *University Library, Cambridge.* **7.9:** Reproduced from *Medicine & the Artist (Ars Medica)*, by permission of *the Philadelphia Museum of Art.* **7.10:** *The Wellcome Trustees.* **7.12:** Reproduced from *Medicine & the Artist (Ars Medica)*, by permission of *the Philadelphia Museum of Art.* **7.13:** *University of Bristol, Faculty of Arts, Photographic Unit.* **7.14:** From *The Microcosm of London.* Reproduced from *Medicine & the Artist (Ars Medica)*, by permission of *the Philadelphia Museum of Art.* **7.15:** BM Egerton Ms. 2572, fol. 51v. *British Museum.* **7.16:** Fitzwilliam Ms. 298, fol. 25. *The Wellcome Trustees.* **7.17:** *British Museum.* **7.18:** *Institut Pasteur.* **7.19:** From Sir Wm. Watson Cheyne, *Antiseptic surgery, its principles, practice, history and results*, London 1882. **7.20:** No. 53 (left) belonging to M. Kangi Sato; no. 52 (right) belonging to M. Seki Hike. Temporary loan to Nippon Bijutsu Token Hozon Kyokai (*Society for Preservation of Japanese Art Swords*). **7.21:** Wallace Collection no. O/1404. *Crown copyright, the Wallace Collection.* **7.23:** From R. A. F. de Réaumur, 'De l'arrangement que prennent les parties des matières métalliques et minérales, lorsqu'après avoir été mises en fusion, elles viennent à se figer', *Mém. Acad. Sci.* 1724. **7.24:** *Robin Godwin, Department of Geology, Bristol.* **7.25:** From P. C. Grignon, 'Mémoire sur des crystallisations métalliques', *Mémoires de physique, sur le fer*, Paris 1775. **7.26:** Wallace Collection no. A 32. Crown copyright, *Trustees of the Wallace Collection.* **7.27:** From C. S. Smith, *History of Metallography.* Chicago 1960. *University of Chicago Press.* **7.28**, **7.29:** *Dr D. Dingley, Bristol University.* **7.30:** From N. de G. Davies & H. R. Hopgood, *The Tomb of Puyemre, Thebes,* Robb de Peyster Tytus Memorial Series II. New York 1922. *NY Metropolitan Museum of Art.* **7.31:** Singer *et al. History of Technology,* vol. 2, fig. 153. Oxford University Press. *ICI Ltd.* **7.32:** Hans Sachs, *Eygentliche Beschriebung alle Stande...*, fol. ii. Frankfurt-am-Main 1568. **7.33:** Singer *et al. History of Technology,* vol. 3, fig. 256. Oxford University Press. *ICI Ltd.* **7.34:** *The Bowater Organisation.*

8.3–8.7: Air Commodore F. Whittle, 'The early history of the Whittle jet propulsion gas turbine', *Proceedings of the Institute of Mechanical Engineers,* **152** (1945) figs. 1, 7, 14, 32, 2. **8.8:** *Deutsches Museum, Munich.* **8.9:** S. G. Hooker, 'The engine scene', *Aeronautical Journal of the Royal Aeronautical Society,* **74** (1970), 3, fig. 7. **8.10:** *Rolls Royce Ltd.* **8.11:** J. Schmookler, 'Economic sources of inventive activity', *J. Economic History,* **22** (1962) 7. **8.14:** *The Sir William Dunn School of Pathology, Oxford.* **8.15:** *Glaxo Laboratories Ltd.* **8.16:** Data from paper by Allan J. Greene & Andrew J. Schnitz Jr. of the Pfizer Corporation, New York, in *The history of penicillin production,* ed. Albert L. Elder, pp. 81–7. Am. Inst. Chem. Eng. Symposium series 100, vol. 66, 1970. **8.22:** From G. Champetier, 'Structure et propriétés des plastomères', in *Quelques aspects généraux de la Science des Macromolécules,* Paris 1955. *Centre National de la Recherche Scientifique.* **8.23:** After *The Universal Encyclopedia of Machines,* translated by C. van Amerongen, p. 374. *Allen & Unwin* 1964. **8.25–8.34:** *Bell System Technical Journal,* **49**, 7 (1970): p. 1261, fig. 4; p. 1254, fig. 7; p. 1264, fig. 5; p. 1286, fig. 6; p. 1270, fig. 9; p. 1353, fig. 12; p. 1373, fig. 12; p. 1397, fig 2; p. 1411, fig. 4; p. 1466, fig. 17. *Bell Telephone System.*

9.1: *National Technical Museum, Prague.* **9.2** *Cliché Observatoire de Paris.* **9.3:** *Archaeological Survey of India, Government of India.* **9.4:** *National Maritime Museum.* **9.5:** *Crown copyright, Science Museum, London.* **9.6:** *Royal Greenwich Observatory.* **9.7:** *National Astronomy and Ionosphere Centre.* **9.9, 9.10:** *University of Cambridge, Cavendish Laboratory.* **9.11–9.14:** *University of California, Lawrence Berkeley Laboratory.* **9.15:** Graph from W. H. K. Panofsky, 'High-energy physics horizons', *Physics Today*, June 1973, p. 24. **9.16:** CERN, *Courier*, **12**, 9 (1972). **9.17:** *Tony Frelo, National Accelerator Laboratory, Batavia.* **9.18:** Louis Rosen, 'Relevance of particle accelerators to national goals', *Science*, **173** (1971) 491, fig. 1. **9.19:** Sir John Cockcroft, *The Organization of Research Establishments*, facing p. 30. Cambridge University Press 1965. **9.20:** From *Nature*, 23 January 1873. **9.22:** *Bell Telephone Laboratories.* **9.24:** *University of Cambridge, Cavendish Laboratory.* **9.26:** *Rutherford High Energy Laboratory.* **9.27:** *CERN.* **9.28:** *Physics Today*, February 1972. **9.29:** *Annual Report* 1967, p. 61, fig. 1. *CERN.* **9.30:** *Rutherford Lab Report* 1971, p. 25. **9.32:** D. J. de Solla Price, *Little Science, Big Science*, p. 88, fig. 19. Columbia 1963.

10.1, 10.2: *Physics in Perspective*, vol. 1, ISBN 0 309 02037 9, Physics Survey Committee 1972; p. 310. *National Academy of Sciences – National Research Council*, Washington, D.C. **10.3:** Annual Report 1970. *CERN.* **10.4–10.6:** OECD, *Reviews of National Science Policy: United States.* Paris 1968. *OECD.* **10.7:** *Physics in Perspective*, vol. 1, Physics Survey Committee 1972. *National Academy of Sciences – National Research Council*, Washington, D.C. **10.8:** B. R. Williams, 'Research and economic growth – what should we expect?' in OECD, *Science, Economic Growth and Government Policy*, Paris 1963. *B. R. Williams, OECD.* **10.9:** From D. J. de Solla Price, 'Nations must publish or perish'. *International Science & Technology*, October 1967. **10.10:** D. J. de Solla Price, 'Measuring the size of science', *Proceedings of Israel Academy of Sciences & Humanities*, **4**, 6 (1969). *D. J. de Solla Price.* **10.11:** B. R. Williams, *Investment & Technology in Growth, Manchester Statistical Society*, 1964. **10.12:** Data from Z. Griliches, 'Research costs and social returns: hybrid corn and related innovations', *J. Political Economy*, October 1958. *University of Chicago Press.* **10.13:** Z. Griliches, 'Hybrid corn and the economics of innovation', *Science*, **132** (1960) 275–80, fig. 3. Copyright 1960 by the *American Association for the Advancement of Science.*

12.1: From S. A. Stouffer *et al. The American Soldier*, Princeton University Press 1949.

13.1: From J. L. Heilbron, *H. G. J. Moseley. The Life & Letters of an English Physicist 1887–1915.* University of California Press 1973. *A. Ludlow-Hewitt.* **13.2:** From *Times History of the War*, vol. 5, 1915. **13.3, 13.4:** *Imperial War Museum.* **13.5, 13.6:** R. M. Page, *The Origin of Radar*: p. 35, fig. 6; plate 1. Doubleday 1962. *R. M. Page.* **13.7:** *Professor R. V. Jones.* **13.8:** R. M. Page, The Origin of Radar, plate vi. Doubleday 1962. *R. M. Page.* **13.9, 13.12:** *Imperial War Museum.* **13.13:** *Crown copyright, Science Museum, London.* **13.14:** Based on L. N. Ridenout, *Radar System Engineering.* MIT 1947. **13.15:** From J. P. Baxter III, *Scientists Against Time.* MIT 1946. **13.16:** *US Army photograph.* **13.17:** *Royal Radar Establishment, Great Malvern* and *Dr Graham Winbolt.* **13.18:** *Crown copyright. By permission of the Controller of Her Majesty's Stationery Office.* **13.19:** Le Pelley in the Christian Science Monitor © 1966 TCSPS. **13.21–13.23:** *US Army photograph.* **13.26:** *Physics Today*, June 1969, p. 69. **13.27:** *Science*, **169** (July 1970) 461. **13.29:** *Science & Public Affairs, The Bulletin of the Atomic Scientists.* © 1967 by *The Educational Foundation for Nuclear Science.* **13.30:** Based on H. Scoville Jr. 'The limitation of offensive weapons', *Scientific American*, **224**, 1 (1971) 22.

INDEX

Huygens, C., 40, 46
hydrogen bomb, 143–5

ideology, 299–301, 347
India, 81, 87, 166, 211, 258, 266, 271–5
indicators, social, 290–1
industrialization, 267, 269, 274, 277, 279, 283:
   of science, 238–9
industry, 176–8, 180–209, 243, 253–5, 264, 269,
   270–1, 274, 277, 288, 308
innovation, 180–209, 263–4, 349
input–output analysis, 287–8
intelligence, 293–9
international corporations, 262, 269, 277
international scientific community, 109, 111,
   271, 337
internationalism, 52, 109, 111, 222, 244, 269,
   280, 306, 335
invention, 13, 26, 28, 76, 139, 141, 164, 174–209,
   271, 277, 303
invisible colleges, 90, 101, 107, 109
IQ, 296–9
Ireland, 213
Israel, 261, 331
Italy, 38, 46, 47, 61, 84, 86, 122, 153, 193, 271

Jai Singh, Maharajah, 211
Japan, 166, 170, 255, 266, 269–71, 310, 331
jet engine, 172, 181–8, 270, 302, 348–9
Jews, 127, 306
Johns Hopkins University, 60
Johnson, S., 1, 180
Josephson, B. D., 120
Joule, J., 25
journalism, scientific, 118, 137
journals, scientific, 56, 65, 92, 99–104, 106, 277,
   280
Judson, W. L., 180–1

Kafka, F., 210
Kelvin, Lord, *see* Thomson, W.
Kepler, 20, 96
Keynes, J. M., 286
Kistiakowsky, G., 331

Lagrange, J. L., 26, 52, 124
Landau, L. D., 122, 143, 338
Langevin, P., 303
Laplace, P. S., 124
Latin, 92
Latin America, 261
Lawrence, E. O., 217–18
leadership, 197
Leeuwenhoek, A., 41

Leibniz, G. W., 44, 50, 92, 124
Leonardo da Vinci, 68–72, 97, 146, 155
letters journals, 92, 108
letters, private, 91–4, 99
libraries, scientific, 100–2, 241, 277
Liebig, J., 131–5
lighthouses, 30
Lindemann, F. A., 140–2, 303, 312
Linnaeus, C., 52, 54
Lippershay, 20
Lister, J., 164
Little Science, 226, 242
Livingstone, M. S., 217–18
locks, 11
logarithms, 23
Los Alamos, 142–3, 350
Lotka's Law, 105
Louis XIV, 40, 50
Lovatt Evans, C. A., 306
Luria, S., 82
Lysenko, T. D., 34, 88–9

magic, 147–8
magneto, 29
magnetron, 316–17
Malpighi, M., 38, 46
Malthus, T., 130–1, 299
Malvern, Royal Radar Establishment, 253,
   319–20, 350
marketing, 181, 192, 208, 264, 293, 343
Marx, K., 281
marxism, 25, 139, 170, 281
materials science, 172–4, 179, 342
mathematics, 8, 23, 81, 92, 120, 124, 286, 287,
   350
matter, properties of, 172
Maxwell, J. C., 31, 62, 138
Mayer, J. R., 103, 104, 122
McNamara, R., 322
mechanical properties of materials, 168, 171,
   172, 201–3
mechanics, 50, 70, 124, 129, 138–9, 303
mechanisms, 13, 42, 44, 70, 139, 176
medicine: benefits, 336; craft, 16, 147, 152,
   303; history, 96, 147–65; profession, 54, 57,
   65, 149, 157, 163–4; research, 38, 46, 61–2,
   154, 250, 289, 306; scientific, 164, 173, 192,
   344; teaching, 148, 153–4, 158, 268–9, 272;
   theories, 30, 151, 161–2, 346
melons, 35
Mendel, G., 33, 57, 88, 102
Mesopotamia, 8
metal working, 13, 165–74, 269
metallography, 171, 172